Key Concepts

STEM

SCIENCE TECHNOLOGY ENGINEERING MATH

PHYSICS

LIGHT AND SOUND

General Editor: Anne O'Daly

BROWN BEAR BOOKS

Published by Brown Bear Books Ltd
4877 N. Circulo Bujia, Tucson, AZ 85718
USA
and
Studio G14, Regent Studios, 1 Thane Villas,
London N7 7PH, UK

Cataloging-in-Publication data available upon request.

ISBN 978-1-83572-079-0 (ALB)
ISBN 978-1-83572-084-4 (paperback)
ISBN 978-1-83572-089-9 (ebook)

Children's Publisher: Anne O'Daly
Design Manager: Keith Davis
Picture Manager: Sophie Mortimer
Consultant: Chris Woodford

Manufactured in the United States of America
CPSIA compliance information: Batch#AG/5665

Picture Credits
Front Cover: Shutterstock: T. Suguliev.
Interior: NASA: 12-13; **Photos.com**: 46-47; **Public Domain**: Femtoquake 34-35, Vlasini Fotografie 26-27, Why? I ask 40-41; **Shutterstock**: Kostiantyn Ablazov 4, Aumm graphixphoto 31, brinkstock 52-53, deepspace 48-49, Fotogenix 1, 32-33, GagliardPhotography 58, Gyvafoto 42-43, hiopex 44-45, Ivanagott 19, Megaflopp 24-25, NASA 8-9, Evan Novostro 39, Plamens Art 55, PrinceOfLove 60-61, RFArrarons 54, Smileus 6-7, 7, Mikhail Varentsov 22, violetblue 56, Vaclav Volrab 20, weedezign 10-11; **U.S. Federal Government**: U.S. Navy/US Navy photographer 50-51.

All other artwork and photography © Brown Bear Books Ltd

Brown Bear Books has made every attempt to contact the copyright holder. If you have any information about omissions please contact licensing@brownbearbooks.co.uk.

STEM

Key Concepts in STEM: Physics describes the key theories in the study of physics and their practical applications. Both light and sound are forms of energy that travel in waves. Light is a type of electromagnetic radiation, but unlike other types of electromagnetic radiation, such as radio waves and x-rays, light waves can be seen. Explanatory diagrams and informative photographs, detailed features on related aspects of the topics covered and the main scientists involved in the advancement of physics, and definitions of key "Science Words" all enhance the coverage. "Try This" sidebars outline experiments that learners can follow as a first step to practical investigations.

Publisher's note to educators and parents: Our editors have carefully reviewed the web sites included at the end of this book to ensure that they are suitable for students. Many web sites change frequently, however, and we cannot guarantee that a site's future contents will continue to meet our high standards of quality and educational value. Be advised that students should be closely supervised whenever they access the Internet.

Contents

PRODUCING LIGHT

A firefly—also called a lightning bug—is a type of beetle that produces a flashing light from its abdomen. The light is produced by a chemical process within the bug's body.

Light is a type of radiation—the only type that can be seen. Light is produced when anything becomes very hot, for example in a candle flame. But there are also cold sources of light, such as the light produced by a firefly.

Flames from a burning fuel such as wax or oil provided people with their earliest sources of light. Candles are made by surrounding a stringlike wick with a cylinder of wax. The heat of the flame melts the wax next to the wick, and the wax burns to produce light. An oil lamp also has a wick that dips into a reservoir of oil such as kerosene. In the candle and the oil lamp, the burning of the fuel is an example of combustion—a chemical reaction in which the fuel combines with oxygen, giving out heat and light in the process.

The first major improvement on wicks came with gas lighting, using flammable coal gas. This gas normally burns with a yellow smoky flame. But by introducing air and adding a mantle, a white light is produced. The mantle is a mesh coated with the oxides of various rare metals, which become incandescent—emitting a bright light—when they are heated by the gas flame.

SCIENCE WORDS

Fluorescent bulb: Also called a fluorescent tube; an electric lamp consisting of a tube containing mercury vapor, with electrodes at each end. Electric current flowing between the electrodes makes the mercury vapor emit ultraviolet light. This strikes the lining of the tube, which is made from phosphor, a substance that gives off a very bright white light.

Incandescence: The emission of light by an object that is heated to white heat.

Incandescent lamp: An electric bulb that has a filament (usually made from tungsten) in a glass globe containing traces of an inert gas such as argon. The electric current heats the filament to incandescence.

Light From Electricity

The earliest form of electric light was the arc lamp. This was developed by the English scientist Humphry Davy (1778–1829) in about 1808. The arc lamp consisted of two carbon rods, called electrodes, with their ends a short distance apart. When the electrodes were connected to a high-voltage supply, a very bright spark (called an arc) formed between the electrodes. More modern arc lights, with metal electrodes, were used in early movie projectors and searchlights, but they are now obsolete.

When an electric current passes along a piece of thin wire, the wire becomes hot. It may become red hot and even white hot before it melts or burns away. In the 1870s, inventors in the USA and Great Britain tried to find ways of making an electric bulb with a filament that would get white hot without it burning away. In 1879, the American Thomas Alva Edison (1847–1931) and Englishman Joseph Swan (1828–1914) independently produced incandescent electric bulbs. As a filament, they used a thin carbon fiber enclosed in a glass vessel from which all the air had been pumped out. Later, incandescent bulbs had a thin piece of tungsten wire as a filament and contained an inert gas—one that does not react chemically—such as argon, rather than a vacuum.

Cold Light

Toward the end of the 19th century, scientists experimented with passing electricity through gases: metal electrodes carried current to and from a glass tube containing gas at low pressure. Neon gas produced a bright orange light, as used in advertising signs. Mercury vapor produced a blue-green light. The inside of a fluorescent tube is coated with a phosphor, which gives off white light when it is illuminated by a mercury-vapor light inside it.

In the natural world, some animals and plants produce light. Fireflies and glowworms are familiar examples. Some deep-sea fish, such as the angler fish, also emit light to attract prey in the deep oceans. This type of light production is known as bioluminescence.

LED (LIGHT-EMITTING DIODE) LIGHT

Incandescent bulbs and fluorescent tubes began to be phased out in the 2020s. Fluorescent tubes were discontinued in the UK in 2023. Many states in the USA and Canada have also stopped making them or are planning to ban them. Both incandescent and fluorescent bulbs have been largely superseded by LEDs, which work by producing light when an electrical current passes through a microchip. This illuminates the tiny light sources, the LEDs, so giving visible light. LED bulbs do not have filaments that burn out, they remain relatively cool, and they use less electricity. The lifespan of an LED bulb is also thousands of hours longer than the life of an incandescent bulb.

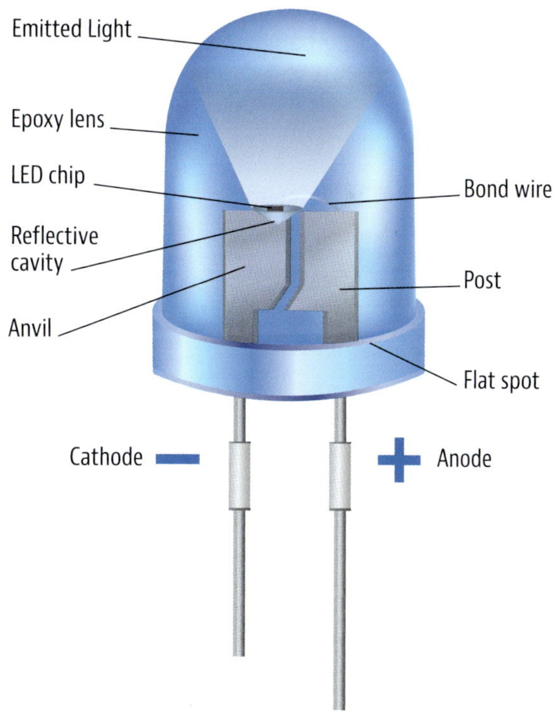

LIGHT AS A FORM OF ENERGY

Energy can be converted from one form into another. The previous pages looked at how chemical reactions and electricity can produce light. Here, we explore how light can be changed into other forms of energy. It enables plants to grow and produces enough electricity to power, for example, a space probe.

The major source of energy on Earth is light from the Sun. Without it no form of life would be able to survive for long. That is because sunlight provides the energy for photosynthesis, the process by which green plants convert carbon dioxide from the air and water from the soil into oxygen and foods such as sugars and starches. Animals are either carnivores (animals that eat other animals) or herbivores (animals that only eat plants). If there was no sunlight, there would be no plants or animals.

In photosynthesis, light energy is converted into chemical energy, which is then stored in sugar and other plant tissues. This is a natural, biological process. However, the conversion of light into electricity involves some quite advanced physics. The simplest form of conversion takes place in a photoelectric cell, such as the type used to make power on some pocket calculators. The key to a photoelectric cell is a substance, such as the semi-metallic element silicon, that emits electrons when light shines on it. The electrons are collected and form an electric current. Photoelectric cells are used to turn streetlights on and off automatically as they respond to the amount of daylight, and are used in burglar alarms.

A field of corn soaks up the sunshine, using the energy of sunlight to convert carbon dioxide and water into sugar and oxygen. The sugar is stored in the plants, while the oxygen passes into the air.

SCIENCE WORDS

Electron: A subatomic particle with a negative electric charge. Electrons surround the nucleus of an atom. They play a key role in electricity, magnetism, and conducting heat.
Photoelectric cell: Also called a photocell, a current-producing device consisting of an element such as silicon that emits electrons when struck by light.
Solar panel: A device that converts light from the Sun into electricity and heat. A solar panel is made from a collection of many small solar cells over a large area that work together to produce large amounts of power from sunlight.

Solar Panels

The current produced by a single photoelectric cell is very small. For larger currents, hundreds of cells are constructed as photovoltaic panels to make electricity. Large solar panels of this type are used to power the communications

and control systems of space probes. This type of solar panel is also used on buildings to create electricity to power homes and offices. Solar panels of a different type can be seen on the roofs of some buildings. These consist of very thin hollow panels containing water, with one of the large surfaces blackened and positioned so that it faces the Sun for most of the day. The blackened surface absorbs solar radiation and heats water that is pumped through the panel. The warmed water may be used in a heating system—it takes less extra energy to heat water that is already warm than to heat cold water.

Solar panels on a roof collect the Sun's light radiation and convert it into electricity for use in the home. Another type of solar panel absorbs heat from the Sun's rays for the home's hot-water system.

POWER IN SPACE

The large "paddles" on a space probe each contain hundreds of photocells. They convert sunlight into electricity to power the probe's electronic systems.

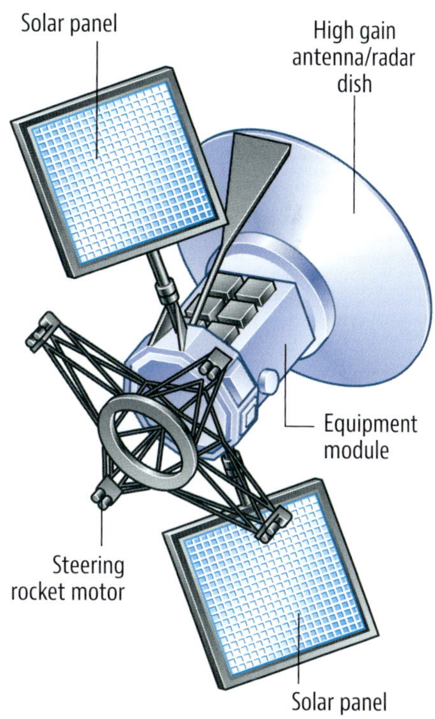

PROPAGATION OF LIGHT

Light from a source such as the Sun or an electric lamp travels out in all directions at an incredibly high speed. It travels in straight lines. Light passes right through transparent substances such as glass and clear plastic. Substances that do not allow light to pass through them are called opaque, and opaque objects cast shadows.

Proving that light travels in straight lines is easy because it makes opaque objects in its way cast shadows. The shadows produced by a small concentrated light source have sharp edges. The shadow is the area that the rays of light from the source cannot reach. The biggest shadow we can ever see is the shadow of the Earth itself. The Sun makes Earth cast a long shadow into space, pointing away from the Sun. Occasionally the Moon moves into Earth's shadow. The Moon shines by reflecting light received from the Sun. However, when Earth's shadow falls on the Moon, the Moon falls into darkness. This is called an eclipse of the Moon, or a lunar eclipse.

Sometimes the Moon, moving in its orbit, passes exactly between Earth and the Sun. A shadow of the Moon tracks across the face of Earth. For anybody in the path of this shadow, the Moon blocks out the light of the Sun, and it becomes nearly as dark as night. This is called an eclipse of the Sun, or a solar eclipse.

Solar eclipses are important to astronomers because they allow them to study the Sun's outer atmosphere, called the corona, which is not normally seen because the Sun is so bright. However, during an eclipse, the bright disk of the Sun is blocked and the corona shows up as pearly light surrounding the dark Moon. The distance between the Sun and the Moon is not always exactly the same. It varies slightly because the Moon's orbit is not perfectly regular. Sometimes the Moon does not completely block out the Sun. (The illustrations on these pages are not to

LUNAR AND SOLAR ECLIPSES

During a solar eclipse **(a)** the Moon passes between Earth and the Sun, stopping the Sun's light from reaching Earth. During a lunar eclipse **(b)** Earth blocks light from the Sun, stopping the Sun's light from reaching the Moon.

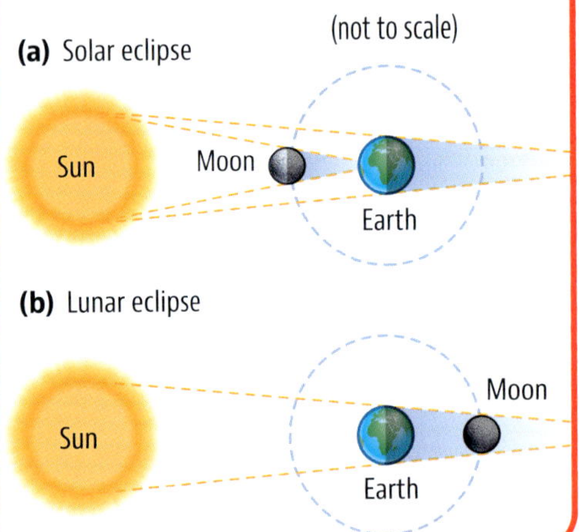

In this image, the Moon is about to block out the light of the Sun, causing a solar eclipse.

scale; the relative distances and sizes are much greater than shown.)

Rays and Beams

Light traveling along its straight path is known as a light ray. (Later in this book we will explore what happens to rays of light when they are reflected by polished surfaces—mirrors—or when they pass through pieces of glass, such as lenses.) A collection or bundle of light rays make up a light beam. Flashlights and searchlights produce beams of light.

Many of the properties of light can be explained by assuming that light travels in waves. For example, the wave theory explains how light is reflected by a mirror or why the colors of the rainbow appear in a soap bubble. In some situations, light behaves as if it is a continuous stream of particles, like a barrage of tiny high-speed bullets from a machine gun. Modern physics can account for both the wave theory and the particle theory of light.

SCIENCE WORDS

Lens: A piece of transparent material that bends light, changing the direction of light rays passing through it.
Lunar eclipse: An eclipse of the Moon, occurring when Earth's shadow (cast by the Sun) falls onto the Moon.
Solar eclipse: An eclipse of the Sun, caused by the Moon passing between the Earth and the Sun.

TOTAL, ANNULAR, AND PARTIAL ECLIPSES

(not to scale)

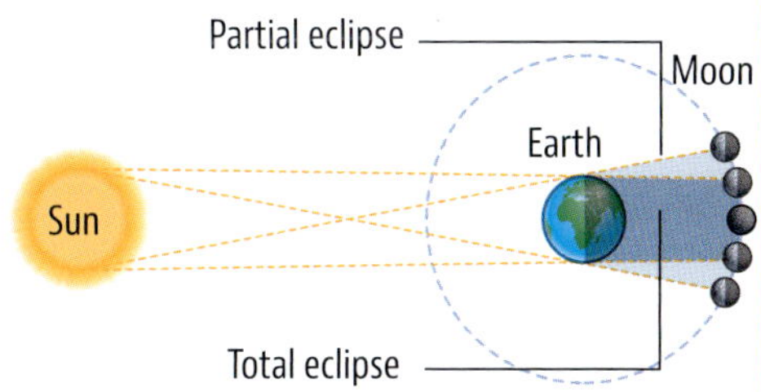

As the Moon orbits Earth, it passes into Earth's shadow, making first a partial eclipse, then a total eclipse of the Moon.

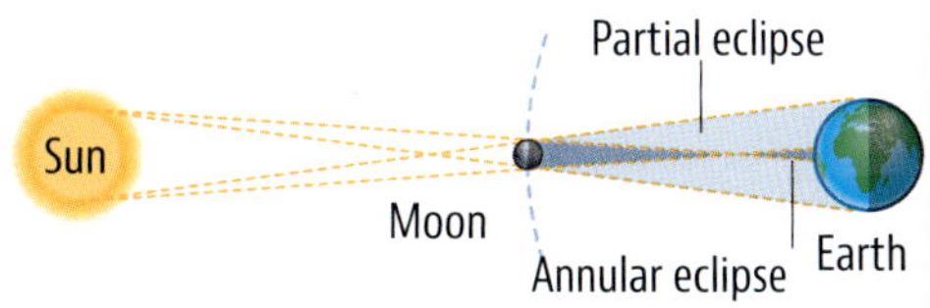

When the Moon is slightly farther away from Earth than usual, it does not completely cover the Sun's disk, making an annular eclipse of the Sun.

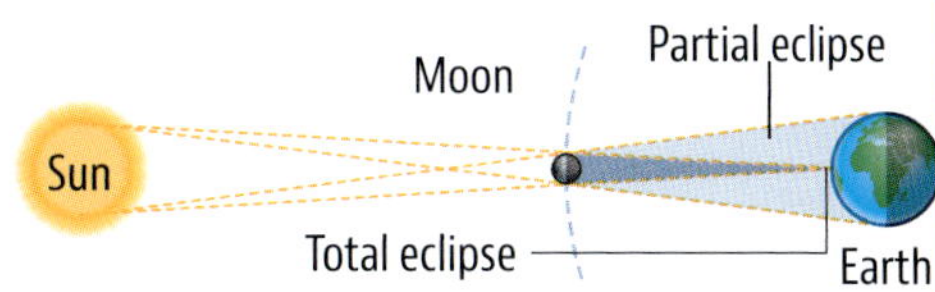

With the Moon at its usual distance from Earth, there is a small region where the solar eclipse is total. Elsewhere it is partial.

SPEED OF LIGHT

Light is the fastest thing in the universe, and nothing can move any faster. It took physicists and astronomers many years to measure the speed of light. This speed is an incredible 186,400 miles per second (300,000 kilometers per second).

When you enter a darkened room and turn on the light switch, the room seems to be flooded with light immediately. In fact, it takes a tiny instant of time for the light to reach your eyes, but light travels so fast it seems to arrive instantly.

The speed of light has been measured as 186,400 miles (300,000 km) per second. At this speed, it takes light reflected from the Moon only just over a second to reach Earth. Light from the Sun has to travel about 93 million miles (150 million km) to reach Earth, and yet it does so in just over 8 minutes.

How Fast Does Light Travel?

For many years, measuring the speed of light proved a challenge to scientists. The first measurement was made by the Danish astronomer Ole Rømer (1644–1710), who in 1676 roughly estimated the speed of light by observing eclipses of Jupiter's moons. In 1690, the Dutch scientist Christiaan Huygens (1629–1695) calculated the speed as just over 143,000 miles (230,000 km) per second, lower than the correct value by nearly 25 percent.

Measurements

More accurate measurements had to await the work of later scientists. In 1849, the French physicist Hippolyte Fizeau (1819–1896) used a rotating toothed wheel to measure the time it took light to make a round trip of 11.2 miles (18 km). His result was within 1 percent of the correct value. Over 30 years later the American scientist Albert A. Michelson (1852–1931) increased the distance traveled by the light to 43.5 miles (70 km). He used rotating mirrors instead of a toothed wheel and obtained a value for the speed of light that was close to the modern figure of 186,400 miles (300,000 km) per second.

Scientific Calculations

In each method, the rotating wheel or mirror drum acted to interrupt the beam of light. The wheel or mirrors were rotated by an electric motor. The observer slowly increased the speed of the motor until the light stopped flickering. The time taken for the light to make the round trip could then be calculated from the rotation speed of the wheel or the mirrors.

SCIENCE WORDS

Prism: A block of glass or plastic that can bend light or split it into the colors of the rainbow. Often triangular in shape.
Reflected ray: A ray of light that is reflected by a mirror.
Vacuum: An empty space in which there are no atoms or molecules of any substance.

Laser beams are often used for entertainment. Scientists decided to direct a laser beam at the Moon to find out how far it was. The beam was reflected back to Earth by a mirror that had been left by Apollo astronauts. As scientists knew the speed of light and the time it took for the beam to make the round trip, it was possible to calculate the distance with accuracy.

Slower Speeds

Light travels through air slightly slower than it travels through a vacuum. Shining a ray of light through a rectangular block of glass slows it down even more. Its speed falls to about 124,000 miles (200,000 km) a second, two-thirds of the speed of light in a vacuum.

Changing Direction

The result of the change in speed is to make the light ray alter its direction inside the glass block. This effect is called refraction, and is described in detail later in the book. The slowing down is caused by the incoming light waves interacting with electrons in the atoms of the glass. As soon as the light ray leaves the glass block, it returns to its original speed and direction. In this way, pieces of glass can bend light rays. This is how lenses and prisms, used in microscopes, binoculars, and other scientific instruments, work.

SPEED OF LIGHT: FIZEAU'S METHOD

Light is reflected by a semisilvered mirror between the teeth of a fast-rotating wheel to another mirror 5.6 miles (9 km) away. The returning beam passes between the next pair of teeth and through the semisilvered mirror to the observer. The wheel speed is adjusted so that there is no flicker of the light when it travels the 5.6 miles (9 km) and back.

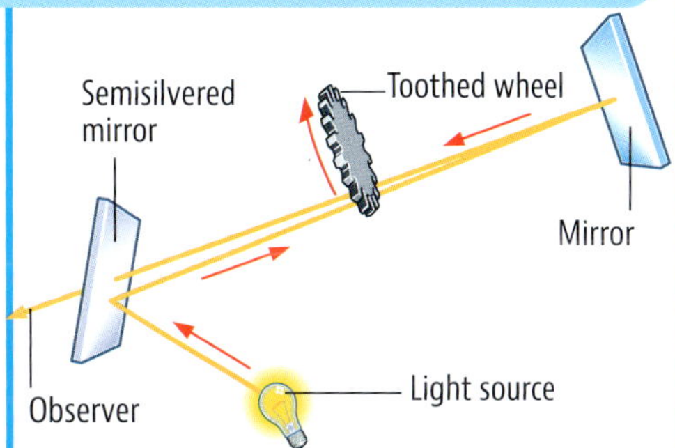

SPEED OF LIGHT: MICHELSON'S METHOD

A rotating drum with a mirror on each side reflects a light beam to a fixed mirror 22 miles (35 km) away. The returning beam is reflected into an eyepiece. The image is steady when the drum rotates by one turn of the mirror drum during the round trip.

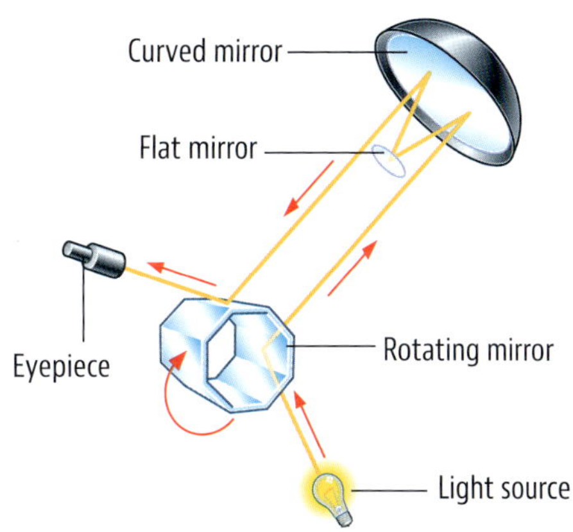

REFLECTION AND REFRACTION OF LIGHT

When a ray of light strikes a flat (plane) mirror, the light is reflected. The light bounces off the mirror at the same angle as it struck it, just like a ball bouncing off the ground. Curved mirrors behave differently, depending on whether they are curved inward (concave) or outward (convex). All types of mirror form images of the objects reflected in them.

All things reflect some of the light that falls on them. If they did not, they would be invisible. Reflected light is scattered in all directions. Flat, or plane, mirrors reflect nearly all the light that falls on them, and they reflect it in the same direction. A ray of light striking a mirror is called the incident ray. The angle at which it strikes the mirror, that is, the angle between the incident ray and a right angle to the mirror (called the normal), is known as the angle of incidence. The angle at which the light ray leaves the mirror is the angle of reflection. According to the laws of reflection of light, for a plane mirror the angle of incidence equals the angle of reflection. The incident ray, the normal, and the reflected ray all lie on the same plane.

SCIENCE WORDS

Angle of incidence: The angle between an incident ray and the normal to a mirror, or to the surface of a block of transparent material.
Angle of reflection: The angle between the reflected ray and the normal to a mirror.
Laws of reflection of light:
1) The angle of incidence equals the angle of reflection.
2) The incident ray, the normal, and the reflected ray all lie in the same plane.
Reflected ray: A ray of light that is reflected by a mirror.

Making Images

When a mirror reflects a light ray from an object, it reaches our eyes. We then look back along the direction of the reflected ray and see an image of the object apparently behind the mirror. It is not a real image—you could not put it on a screen located behind the mirror. It is therefore known as a virtual image. An image that can be put on a screen is called a real image.

Another property of plane (flat) mirrors is that they form same-sized images of objects. The image appears to be the same distance behind the mirror as it actually is in front of the mirror. But if you look at your reflection in a mirror, it seems to be reversed "sideways." If you raise your right arm, the reflected version of you in the mirror seems to raise its left arm. In other words, the

NASA's James Webb Space Telescope, launched in 2021, has 18 hexagonal gold-plated Beryllium mirror segments that combine to make a mirror 21 feet (6.5 m) in diameter.

left-right of your image appears reversed. Physicists call this effect lateral inversion, but it's really an illusion that happens because we have to turn to face a mirror. Mirrors do reverse things, but front to back rather than side to side.

Plane Mirrors

The commonest use of plane mirrors is for looking at our own reflection. Every day people use a mirror when combing their hair, when putting on makeup, or when shaving. Clothes stores have long mirrors for their customers. Mirrors are also used to add light and a feeling of space to rooms. A carefully placed mirror is a good way of getting reflected light into a dark room or corridor to improve visibility. And because it is not always obvious that we are looking at a reflection in a mirror, and not at a real object, illusionists and magicians employ mirrors for some of their on-stage trickery. Periscopes also use plane mirrors, angled at 45°. Periscopes are used to look over obstructions, especially if you are not tall enough to see over the heads of people in front of you at a parade or sporting event. Enlarging periscopes that are used in submarines usually contain prisms instead of mirrors.

Curved Mirrors

Curved mirrors behave in quite a different way. There are two main types, called concave if they are curved inward like the inside of the bowl of a spoon, and convex if they are curved outward like the outside of the bowl of the spoon. The curvature gives such mirrors two further properties. Each has an axis, which is a line at right angles to the mirror that passes through its center. And the radius of curvature is the distance to the mirror from the center of a sphere of which it would form a part. The center of this sphere is also the mirror's center of curvature.

LAWS OF REFLECTION

At a plane mirror the angles of incidence and reflection are equal. The incident ray, normal, and reflected ray are in the same plane.

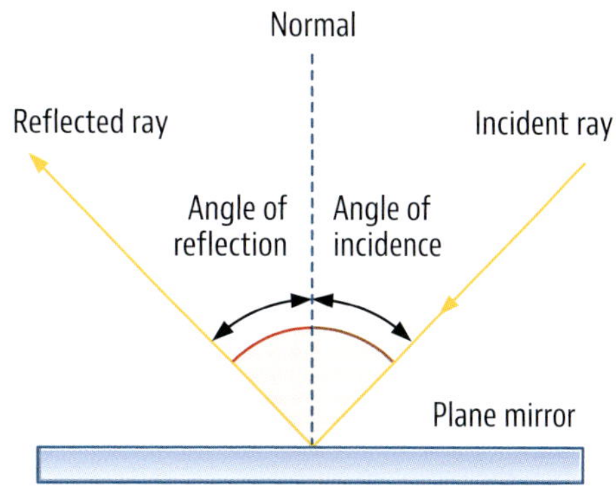

TRY THIS

Multiple Reflections

A single mirror produces a single reflection—an image—of an object placed in front of it. Two mirrors will provide two images and effectively let you see around corners. This project creates multiple reflections from more than one mirror.

What to do

Put two mirrors together with their reflecting surfaces facing each other. Join them down one side with masking tape or Scotch tape. Stand the mirrors on the sheet of paper as shown in the illustration below left. Place a coin in front of the mirrors. How many reflections can you see?

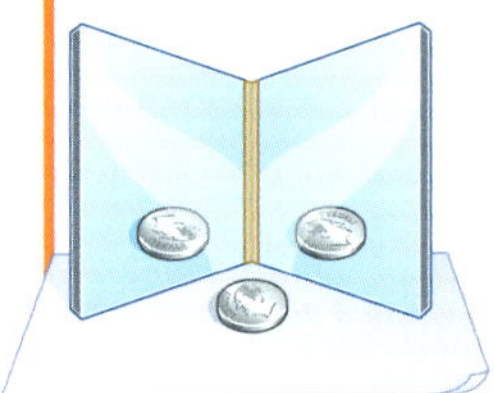

Look carefully in each mirror at the reflection of the coin. Are they exactly the same?

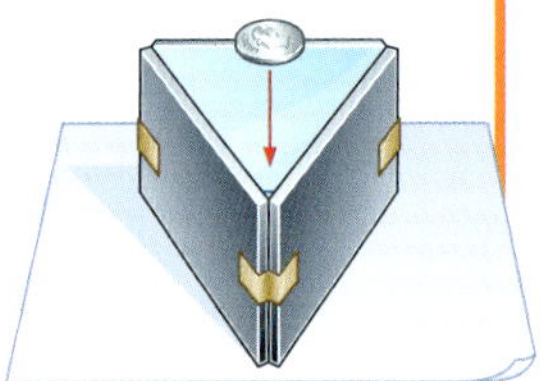

Reflections are more complex if you use three mirrors with a coin in the center.

Change the coin for a small toy such as a car or a doll. Notice how you can see the front of the object in one mirror and the side of the object in the other mirror. You are, in effect, seeing around a corner. What happens to the images if you change the angle between the mirrors by "folding" them closer together and "unfolding" them farther apart? Draw a line across the paper, put the folded mirrors across it, and watch how its reflections change as you slightly open and close the mirrors. If you have three mirrors, tape them into a triangle with a coin in the middle. How many reflections can you see?

With a concave mirror, rays of light parallel to the axis are reflected to a point known as the focus. For this reason a concave mirror is also called a converging mirror. When a convex mirror reflects parallel rays, however, the reflected rays fan out to form a diverging beam. These rays all appear to come from a single point behind the mirror, which is its focus. A convex mirror is therefore also known as a diverging mirror. The concave mirror has a real focus, while the convex mirror has a virtual focus. In both types of mirror, the focal length—the distance from the mirror to the focus—is half the mirror's radius of curvature.

Images in Curved Mirrors

The formation of images by curved mirrors is more complicated than with plane mirrors. What happens depends on whether the mirror is concave or convex, and on how far the object is from the mirror. For a concave mirror, there are four different possibilities.

When the object is farther from the mirror than the center of curvature, the image is upside down, smaller than the object, and located in front of the mirror. You can see this for yourself by looking into the bowl of a polished tablespoon from just a few inches away.

When the object is moved closer to the concave mirror's center of curvature, it grows in size until, at exactly the center of curvature, the image (still inverted) is the same size as the object.

When the object is even closer to the mirror, between the center of curvature and the focus, the image is still real and inverted but it is now larger than the object. The mirror now magnifies.

Finally, when the object is between the focus and the mirror, the image becomes virtual (is formed behind the mirror),

SNELL'S LAW

Snell's law, the chief law of refraction, states that the sine of the angle of incidence divided by the sine of the angle of refraction is a constant, known as the refractive index.

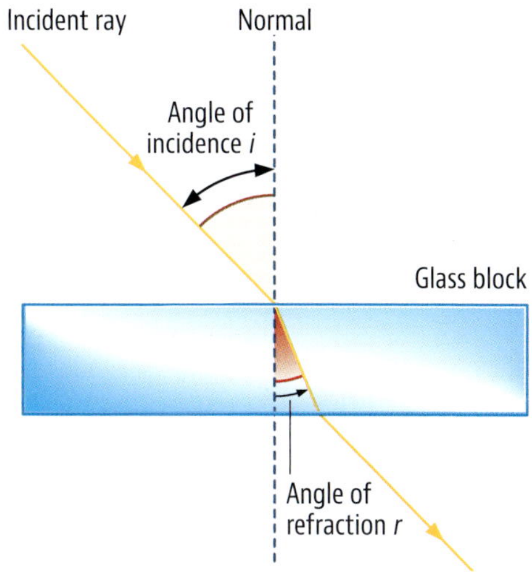

Refractive index = $\frac{\sin i}{\sin r}$

magnified, and the right way up. Such magnified images can be seen in mirrors designed to be used when shaving or putting on makeup.

A convex mirror always produces a reduced, upright virtual image. It is the type used as a driving mirror for motor vehicles. Because the whole scene is reduced in size, it provides a wide angle of view. Both types of curved mirror are used in telescopes (see page 28).

Refraction of Light

A swimming pool often looks as if it is not as deep as it really is. Fish in a lake or river often appear to be nearer the surface than they really are. The reason for this is that light rays traveling from underwater objects do not keep going in the same direction when they come through the surface of water into the air. A similar effect occurs when light rays pass from air into water. The angle between the incoming ray and the normal (a line at right angles to the surface) is called the angle of incidence. Below the water surface, the angle between the light ray and the normal is called the angle of refraction. When light enters a denser medium, as when it travels from air into water or into glass, the angle of refraction is less than the angle of incidence—the ray is refracted toward the normal. When light travels from one medium into a less dense medium, as from glass into air, the angle of refraction is greater than the angle of incidence—the ray is refracted away from the normal.

As with the reflection of light, there are laws of refraction. The laws concern the angles—not the angles themselves, but a mathematical function called the sine (usually written as sin) of the angle. The chief law says that the sine of the angle of incidence (sin *i*) divided by the sine of the angle of refraction (sin *r*) has a constant value for any pair of media. This ratio is known as the refractive index. For air to glass it is about 1.5, and for

SCIENCE WORDS

Concave mirror: Also called a converging mirror, a type of mirror that causes parallel rays of light to be reflected to a focus in front of the mirror. Its surface curves inward.
Convex mirror: Also called a diverging mirror, a type of mirror that causes parallel rays of light to spread out (diverge) after reflection so that they appear to come from a point (the focus) behind the mirror. Its surface curves outward.
Refraction: The bending of light rays as they pass from one transparent material into another.

BENT STRAW

Refraction causes a straw to appear bent below the surface of water in a glass. The same effect makes a pool appear to be shallower than it really is, and for fish to look as if they are nearer the surface of the water than they actually are.

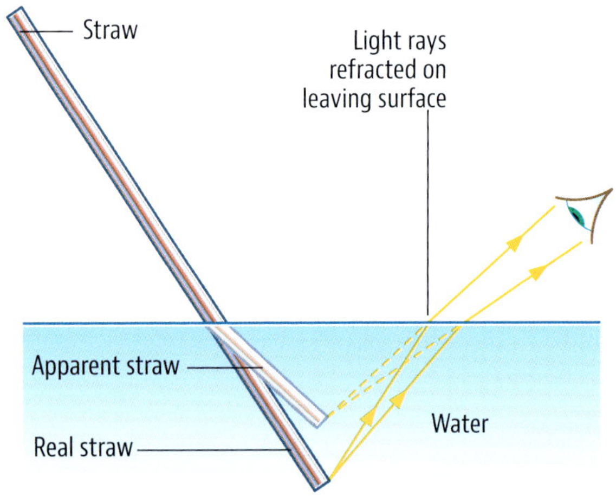

air to water it is about 1.33. The law is also known as Snell's Law, after the Dutch physicist who formulated it nearly 400 years ago.

WILLEBRORD SNELL

Willebrord van Roijen Snell was born in 1580 at Leiden in the Netherlands. He trained in mathematics and physics and when his father died in 1613, Willebrord succeeded him as Professor of Mathematics at the new Leiden University. Snell specialized in land measurement and mapping, and carried out many experiments on light and optics. He discovered his law of refraction in 1621 and introduced the idea of refractive index (now defined as the ratio of the sines of the angles of incidence and refraction). When Snell died in 1626, the results of his work had still not been published. It was later found that the refractive index is also equal to the ratio of the speed of light in the two media concerned.

The second law of refraction states that the incident ray, the normal, and the refracted ray all lie in the same plane (just as with reflection—see page 13).

Bent Straws and Sunbeams

Refraction can produce some strange effects. If you look at a drinking straw placed in a glass of water, the straw appears to bend below the surface. That is because light rays traveling from the straw and leaving the surface are refracted away from the normal. When we look back along the emerging rays, we see the end of the straw at a position that is apparently nearer the surface (as in the illustration above).

A similar effect can occur with the setting Sun, when the air near the surface is denser than that above it. Light rays from the Sun are refracted as they pass through this denser air. Again, looking back along the refracted rays, we see the Sun in a different position. As a result, we appear to be able to see the Sun even when it has dropped below the horizon.

In the opposite situation, when light is traveling from dense air through less dense air, refraction also occurs, and a mirage can be the result. In the case of a mirage, the warm air near the ground is less dense than the colder air above it, a condition that often arises in deserts and above the surface of a highway in warm weather. Light rays from a distant object follow a curved path through the warm air. When we look back along these rays, we see an image of the distant object, but the image is upside down and appears to be below the ground surface.

Critical Angle

When the angle of incidence reaches a certain value, which physicists call the critical angle, the angle of refraction equals 90°. In other words, the refracted light ray travels along the boundary between the two media. This is called total internal reflection. If the angle of incidence is greater then the critical angle, there is no refraction. The incident ray is then reflected from the surface of the second medium, just as if it had struck a mirror.

Uses of Refraction

The major practical use of refraction is in lenses and in the many different types of instruments and devices that employ them. These include cameras, microscopes, telescopes, binoculars, and projectors (see pages 26 to 31). Optical prisms also make use of refraction (see pages 18 to 19).

Total internal reflection is also the physics behind fiber optics, in which light is "piped" along a bundle of thin fibers of glass or plastic. A succession of refractions and internal reflections occur along the length of each fiber so that most of the light entering at one end comes out at the other end. Fiber optics are used in medical endoscopes for making examinations inside a patient's body and for long-distance telephone cables in which signals are transmitted as a series of coded flashes of light. Many thousands of telephone conversations can be sent at the same time along a single optic fiber.

TRY THIS

Hidden Money

When light rays pass from one transparent medium into another, they are bent. This process is called refraction. In this project, you will use refraction to make a coin appear.

What to do

Put a large, shallow bowl on a table, and place a coin on the bottom of the bowl between the center and the edge nearest to you. Back off slowly, looking at the coin, until the coin is out of sight behind the edge of the bowl. Ask a friend to slowly fill the bowl with water. The coin comes back into view. Water refracts light rays from the coin. As the rays leave the surface of the water, they are bent toward you. This is why fish in a pond always look nearer the surface. Birds have to take this into account and "aim off" when trying to grab their prey.

Stand where you cannot see the coin in the empty bowl.

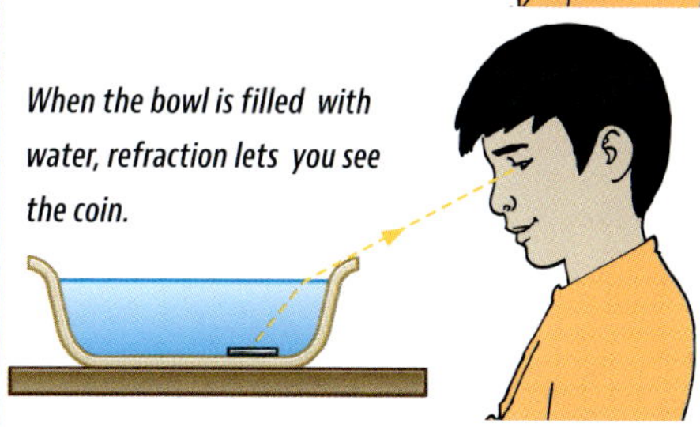

When the bowl is filled with water, refraction lets you see the coin.

PRISMS AND LENSES

Prisms are the best-known light benders. Refraction bends a light ray when it enters a prism and then bends it again when it leaves. More importantly, it bends different colors of light to a different extent. The different colors produced are called a spectrum. A triangular prism can split white light from the Sun into all the colors of the rainbow—called the solar spectrum.

SCIENCE WORDS

Dispersion: The splitting of white light into the colors of the rainbow (a spectrum), for example, by a triangular prism. Raindrops cause dispersion in a rainbow.
Prism: A block of glass or plastic that can bend light or split it into the colors of the rainbow. Often triangular in shape.
Refracted ray: A ray of light that is refracted as it passes from one transparent material into another.

One of the most important experiments in physics took place in a darkened room in Cambridge, England, in about 1665. The physicist Isaac Newton (1643–1727) allowed a beam of sunlight through a hole in the drapes and shone it onto a glass prism. To his surprise, parallel bands of rainbow colors appeared on the opposite wall. From this observation Newton concluded that sunlight consists of a mixture of colors that the prism had separated. When he selected just one of the colors and passed it through a second prism, there was no further change. Modern physics can easily explain what happened in Newton's room. White light is made up of a mixture of the colors of the rainbow, ranging from red to violet with all the other colors in between. As each color enters the prism it is

SINGLE-LENS REFLEX

There are two types of reflector in a modern single-lens reflex (SLR) camera: a mirror and a five-sided prism called a pentaprism. Light entering the camera through the lens is first reflected upward by the mirror. Then two more reflections inside the pentaprism direct the light out of the viewfinder into the photographer's eye. A pentaprism is used because the image leaving the lens is upside down, and the double reflection in the pentaprism turns the image right side up again. When you press the shutter, the mirror flips out of the way and light hits a sensor behind it to capture the image.

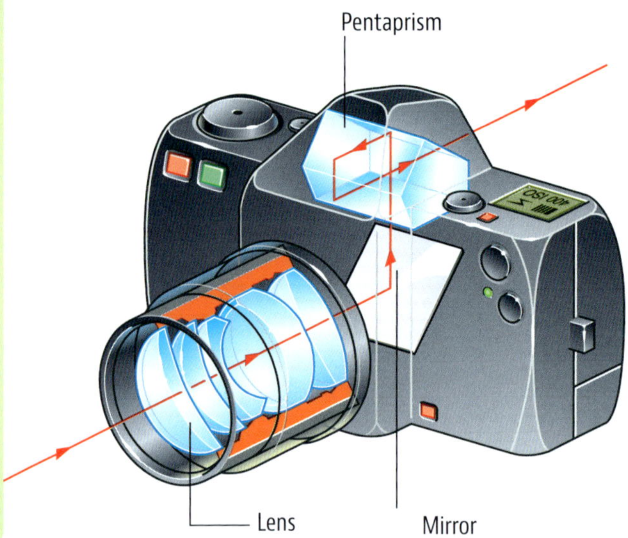

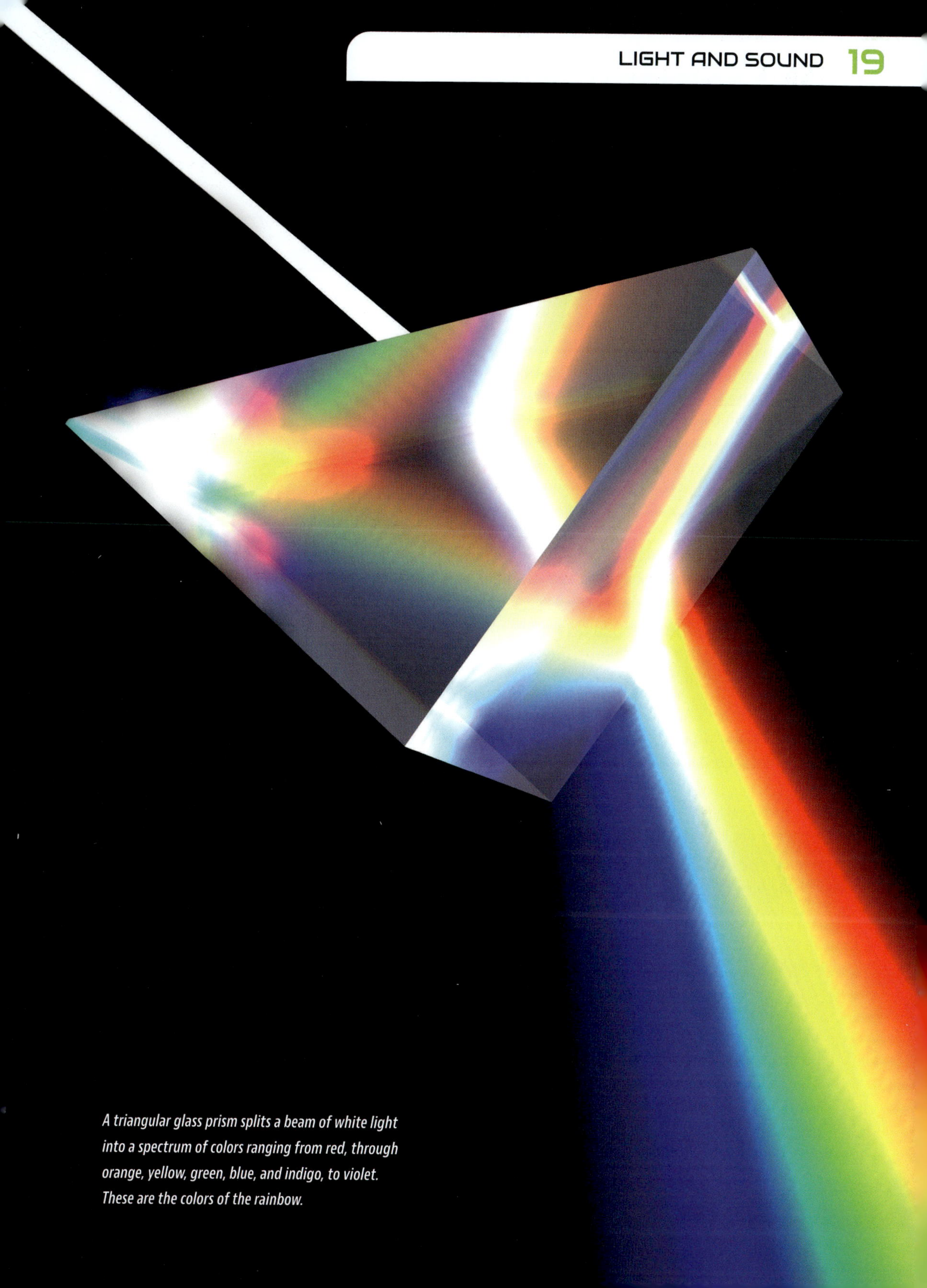

A triangular glass prism splits a beam of white light into a spectrum of colors ranging from red, through orange, yellow, green, blue, and indigo, to violet. These are the colors of the rainbow.

refracted (bent). Red light is not bent as much as violet light. As a result, the red and violet emerge from the prism at different angles (the in-between colors emerge at in-between angles). This has the effect of spreading white light's component colors into a spectrum: red, orange, yellow, green, blue, indigo, and violet.

This special sort of refraction by a prism is known as dispersion, and the different colors produced are called a spectrum. This accounts for the colors that can sometimes been seen when sunlight shines through a crystal glass or ornamental light fitting. It also accounts for the formation of rainbows (see page 22).

Uses of Prisms

Prisms are used in several scientific instruments, such as spectrographs, as well as in periscopes and binoculars. Perhaps the most common use for prisms today is in the single-lens reflex (SLR) camera. Everyday prisms, such as the one Isaac Newton used, are triangular in shape. However, the SLR camera makes use of a prism that has five faces. This is called a pentaprism.

A water droplet viewed through a magnifying glass. The lenses used in magnifying glasses are convex or converging lenses.

Lenses

There are two main kinds of lens. A lens where the surfaces bulge outward is called a convex lens. This is the type of lens that is used as a magnifying glass. Because parallel light rays that pass through a convex lens converge (come together) to a focus on the other side of the lens, this type of lens is commonly referred to as a converging lens.

A lens where the surfaces are curved inward is called a concave lens. It is the

CONVERGING AND DIVERGING

The convex lens is also called a converging lens because parallel light rays passing through it converge (come together) to a focus.

The light rays diverge (spread out) on passing through a concave lens, which is also called a diverging lens.

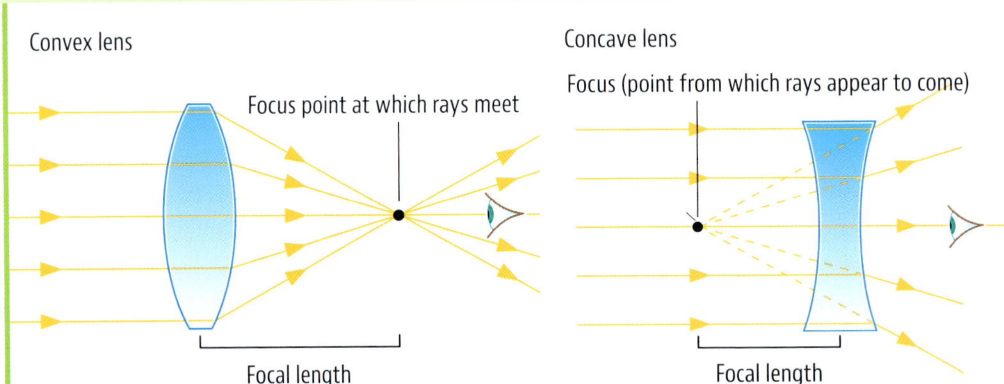

SCIENCE WORDS

Convex lens: Also called a converging lens, a type of lens that causes parallel rays of light to converge to a point (the focus) in front of the lens. Its surfaces curve outward.
Concave lens: Also called a diverging lens, a type of lens that causes parallel rays of light to spread out (diverge) as if coming from a point (the focus) behind the lens. Its surfaces curve inward.

type of lens used in eyeglasses for shortsighted people. Parallel light rays pass through a concave lens and diverge (spread out). The focus is on the same side as the incident light. For this reason, a concave lens is also referred to as a diverging lens.

Larger and Smaller

It has already been noted that a convex lens can be used as a magnifying glass. But how exactly does this work? Parallel light rays from an object converge toward a focus, and when we look back along these rays, we see a magnified image of the object. With a concave lens the light rays diverge toward the observer's eye. Looking back along these rays reveals a diminished image of the object. Artists and designers sometimes use a diverging lens, which they call a reducing glass, to check how a large image will appear when it is reduced in size.

Faulty Lenses

If you look at an object through a simple convex lens, you can often see colored fringes around the edge of the image. This effect is called chromatic aberration (any fault in a lens is known as an aberration). This happens because the edges of the lens refract blue light more than red light so that the two colors are focused at different points.

TRY THIS

Reversing Arrow

This very simple project can be proposed in the form of a puzzle: how can you reverse the direction of an arrow without touching it?

What to do

Use a colored pen to draw a fat horizontal arrow pointing to the left on a piece of cardboard. Prop the cardboard upright against a drinking glass. How can you reverse the arrow without touching the paper?

Fill another glass with water and stand it several inches in front of the cardboard. Look at the arrow. It is now pointing to the right! What happens is that the glass of water acts like a fat lens and produces a reversed image of the arrow. Place the glass of water nearer the paper, and the arrow will go back to pointing to the left.

Stand a glass of water in front of the arrow to reverse its direction.

Color Correction

The effects of chromatic aberration can be corrected by adding a concave lens that is made from a different kind of glass. This makes the blue rays diverge and come to the same focus as the red rays. The resulting combination is called an achromatic lens (or achromat).

LIGHT AND COLOR

A bright rainbow arches down. A faint secondary rainbow can be seen to the right. Its colors are in the opposite order.

White light is a mixture of all the colors of the rainbow. So why do colored objects appear colored when they are illuminated with white light? And why do most colored objects change color when illuminated with colored light?

The illustration below explains why colored objects look colored. When white light illuminates a red object, for example, most of the light is absorbed by the surface of the object. Most—but not all. The red component of the white light is reflected, along with a little orange. As a result, the object appears to our eyes to be red. Similarly, a blue object reflects mostly blue light, and a yellow object reflects mostly yellow.

This explanation works only for white light. If objects are Illuminated with colored light it can have strange effects. Try looking at the colors of cars and trucks under old-style yellow sodium streetlights, and you will not be able to get the colors right. In fact, only yellow vehicles will still look yellow.

Mixing Colored Lights

White light is a mixture of the seven colors of the rainbow. So it should come as no surprise to learn that mixing all seven colors in the right amounts produces white light. But not all seven colors are needed. You can make white light by mixing only three colors, which are known as the primary colors of light. Light's primary colors are red, blue, and green. All three mixed together make white. The primary colors can be mixed to make three more colors, called secondary colors. Red and green light mix to make yellow; green and blue light mix to make a bright blue called cyan; blue and red mix to make a pinkish red called magenta.

Making Pictures

A mixture of the three secondary colors again produces white light. In fact, careful mixing of the three primary or secondary colors of light

COLORED OBJECTS

White light is a mixture of all colors. When it shines on a colored object, the object absorbs all colors except its own. This color is reflected to the eyes of the observer.

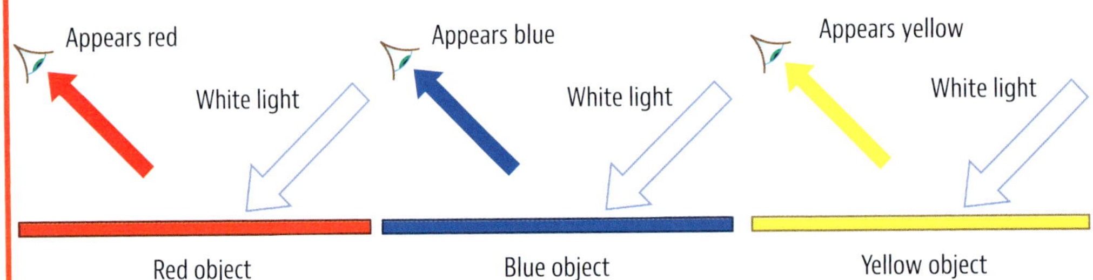

will produce any color you can think of. If you look at a television picture up real close, you will see that it is made up from lots of tiny colored dots. Look even closer, and you will see that there are only three kinds of colored dot: red, green, and blue—the primary colors. In this way, the correct mixture of dots produces the complete range of colors on the TV screen. The colors of a color transparency photograph are produced in much the same way. Because the colors are added to make new colors, this type of color mixing is called the additive process.

Mixing Colored Paints

So far we have looked at the mixing of colored lights. More familiar to most people is the effect of mixing paints, inks, and other pigments. If you mix together all the colors in a paintbox, you will get a muddy black color.

As with colored lights, colored paints have three primary colors: yellow, cyan, and magenta (exactly the same as the secondary colors of light). A mixture of all three primary colors produces black. Mixing the primary colors in pairs produces the secondary colors of paint: yellow and magenta mix to make red, magenta and cyan mix to make blue, and cyan and yellow mix to produce green. Notice that the secondary colors of paint are the same as the primary colors of light.

When yellow and magenta paints are mixed, this has the effect of removing blue and green from black, which leaves red. Similarly, mixing magenta and cyan removes red and green to leave blue, and mixing cyan and yellow removes red and blue to leave green.

Artists learn how to make other colors by mixing. In fact, some artists make most of the colors they need by combining them. Because mixing colored paints effectively removes colors from black, it is called the subtractive process.

TRY THIS

Wheel of Colors

This project shows you how to combine all the colors of the rainbow to make white.

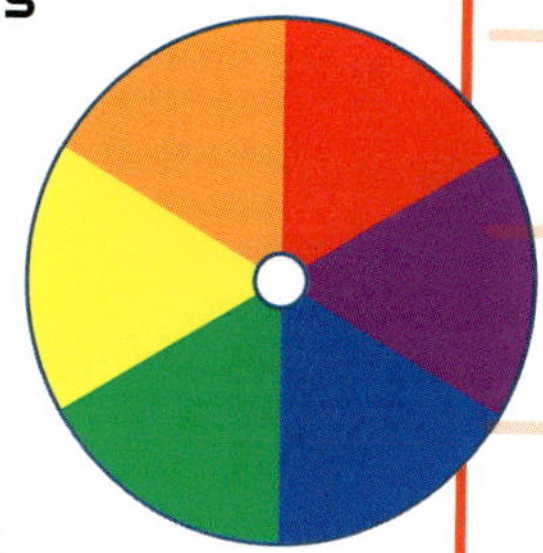

Color the six segments as shown above.

What to do

Divide a disk of cardboard into six segments. Each segment should make an angle of 60° at the center. If you do not have a protractor, draw the segments by eye, looking at the above illustration to see what the angles are like; they do not have to be exact.

Color the segments as shown, using paints or crayons. The colors you need are red, orange, yellow, green, blue, and violet. Carefully push the point of a pencil through the center of the disk to make a kind of top. Spin the top. As it spins around, what color do you see? You have recombined the rainbow colors to make white.

Try the same experiment using other colors to make yet more colors.

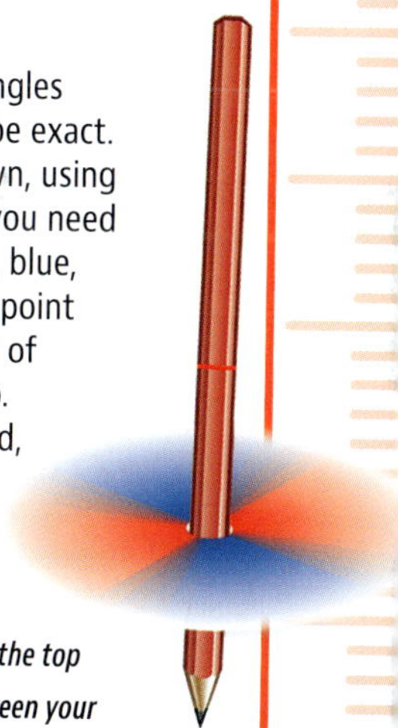

Spin the top between your thumb and first finger; this takes a little practice.

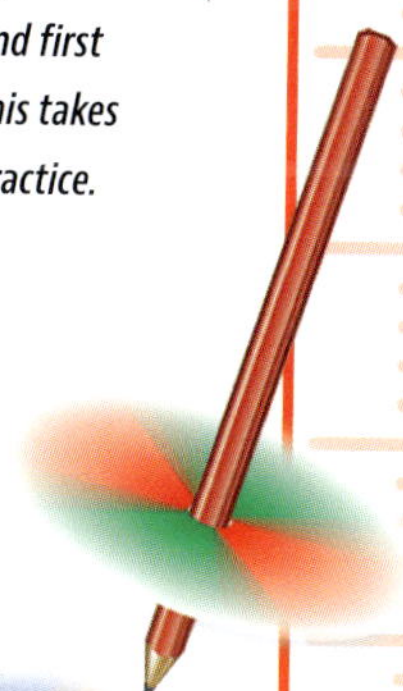

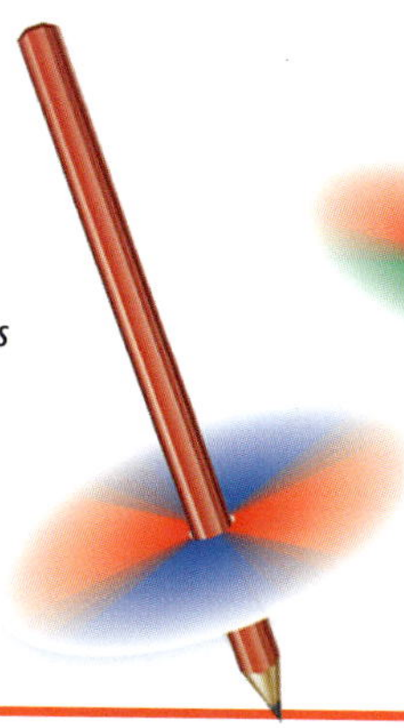

Make two more disks with different sets of colors.

THE HUMAN EYE

Light has been defined as the only type of radiation that we can see, but we could not see anything if we did not have eyes. The human eye is a natural application of a lens. Knowing how lenses work gives us an understanding of the human eye, some of the defects of human vision, and how to correct them.

The main parts of the human eye are shown in the illustration below. The lens and its supporting structures divide the eyeball into two chambers. The front chamber contains a watery liquid called aqueous humor. The jellylike vitreous substance fills the chamber. The eyeball itself is transparent at the front, the cornea, so that light can enter. A thin layer of tears keeps the cornea moist.

ANATOMY OF THE EYE

Important parts of the eye include the lens and the cornea. Both help in focusing, although only the lens is adjustable—it does most of the work.

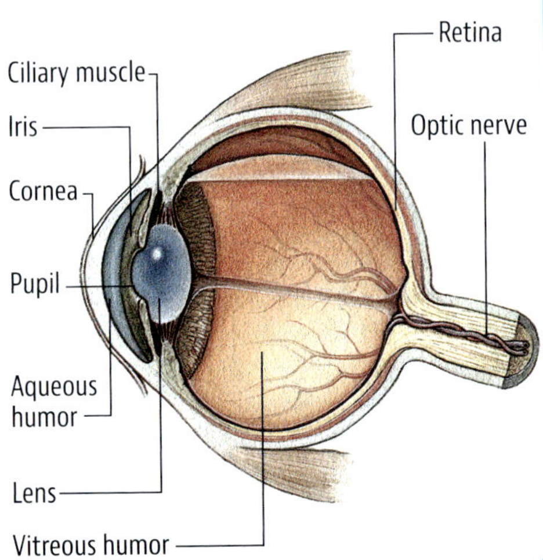

The lens is supported by ciliary muscles, which also pull on the lens to change its shape to focus on objects. The lens is stretched and made thinner by the muscles when we look at distant objects. The lens is relaxed and made thicker when we look at nearby objects. The colored iris in front of the lens has a central hole called the pupil. The iris changes size to vary the size of the pupil, which is large in dim light in order to let in as much light as possible, but much smaller in bright light to protect the eye and keep light out.

Light Detection

The light-sensitive retina lines the inside of the eyeball, and the lens focuses an upside-down image of objects onto the retina. There the light triggers nerve impulses, which pass along the optic nerve to the brain. The brain then combines the impulses from both eyes, converts them into "pictures" that we can see, and turns them the right way up.

Eye Defects

The two upper diagrams on the right show a normal eye and how it forms images. In two of the most common defects of eyesight, light rays are not focused correctly on the retina. In a longsighted person, the eyeball is too short

The colored part of the eye is the iris, in front of the black pupil. Light from everything that we look at passes through the pupil and into the eyeball.

front-to-back. As a result, the eye's lens tries to focus light rays at a point behind the retina. This condition is corrected with eyeglasses made from convex lenses, or by convex contact lenses. They make the rays converge in focus onto the retina.

In a shortsighted person, the eyeball is too long front-to-back. The eye's lens brings light rays to a focus that lies in front of the retina. To improve their vision, shortsighted people wear eyeglasses made from concave lenses, or concave contact lenses. They make the rays diverge slightly, so that they come to the right focus on the retina.

Another common eye defect, called astigmatism, arises when the transparent cornea at the front of the eyeball is not perfectly spherical. When looking at a cross like a plus sign (+) with an astigmatic eye, either the upright part of the cross is in focus and the horizontal part is not, or vice versa. The condition is corrected with eyeglasses or contact lenses that have a similar fault, but at right angles to that of the eye. Such lenses are called anastigmatic.

VISION IN FOCUS

In a normal human eye, the lens produces an image that is much smaller than actual size. In a longsighted eye, light rays are focused behind the retina. This can be corrected with convex lenses. In a shortsighted eye, the rays are focused in front of the retina; concave lenses correct this condition.

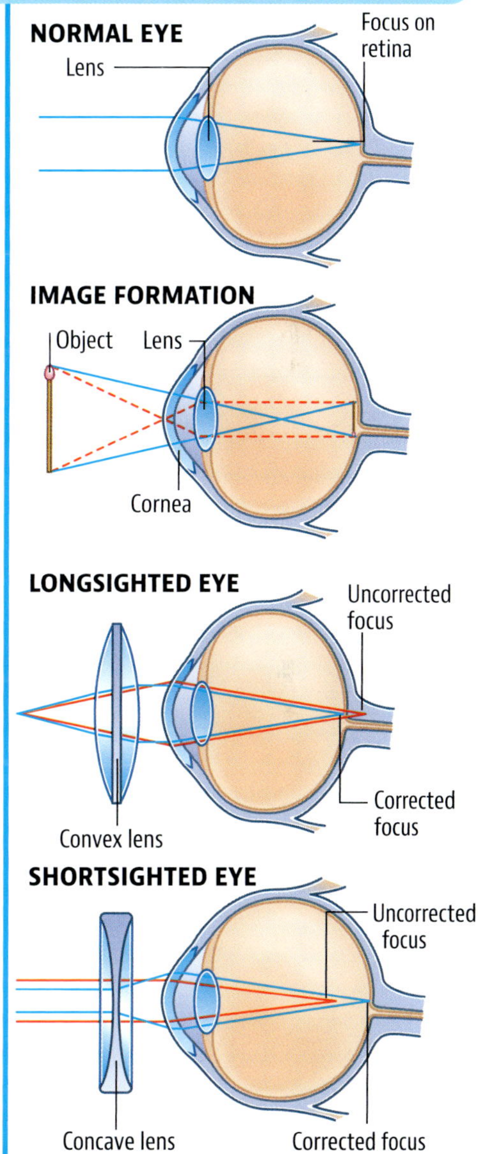

USES OF LIGHT

The main components of optical devices are mirrors, lenses, and prisms. The ways they work have been dealt with earlier in this book. Now we look at how such devices have been put to practical use in the development of various kinds of optical instruments.

Today, nearly everyone owns at least one camera. Although most are cameras built into cell phones, some are standalone digital cameras, and a few are analog, or non-electronic. In many ways a camera works like the human eye. It has a lens, a variable aperture, and a light-sensitive surface. The lens focuses the image, the aperture controls the amount of light entering the camera, and the film records the image focused onto it by the lens. The image is actually upside down, but this does not matter. In addition, a camera has a shutter that controls the amount of time (in fractions of a second) for which light is allowed into the camera and onto the film.

In order to focus the object being photographed—depending on how far away it is—the lens can be moved farther from or nearer to the film. On a cell-phone camera there are often several lenses side by side, and the phone simply switches to a different lens automatically. On an analog or digital camera, focus is usually achieved by rotating the lens mount, which has a fine screw thread. In very old cameras, as well as some specialized modern ones, the lens is mounted on a bellows arrangement and racks backward or forward to focus the camera. In this respect, the camera is different from the human eye, in which the lens changes shape in order to focus. But there are animals that focus their eyes just like a camera. An octopus is able to move the lens of its eye in and out. Cameras have a device called a viewfinder

SIMPLE CAMERA

A simple analog film camera is basically a lightproof box holding light-sensitive film. A lens focuses an image of the object to be photographed onto the film. The lens can be screwed in or out to focus the image. An adjustable aperture, called an iris diaphragm, can be altered in size to control the amount of light entering the camera. In a simple digital camera, the film is replaced by a light sensor.

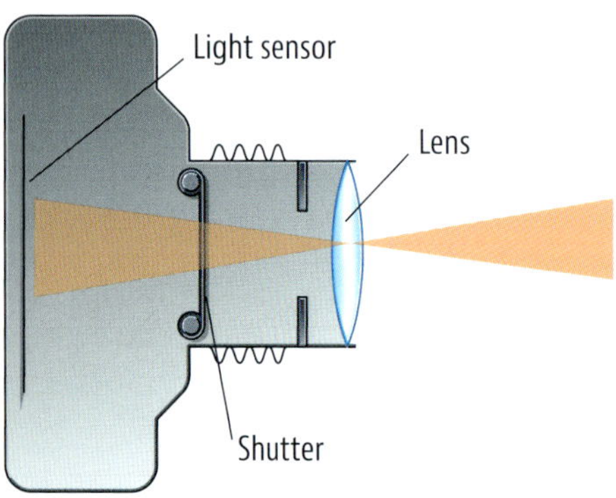

A 6.5 foot (2 m) reflecting telescope at an observatory. The figure refers to the diameter of the primary mirror.

that allows the photographer to aim the camera accurately. The viewfinder of a simple camera is a pair of small lenses. In a single-lens reflex camera a pentaprism forms the viewfinder (see page 18). Some cameras have interchangeable lenses—alternative lenses for different tasks. They range from wide-angle lenses with short focal lengths to long-focus telephoto lenses for taking closeups of distant objects.

Digital cameras have dispensed with film altogether and use computer technology to record images, but they still rely on lenses, prisms, and mirrors to form the image.

Objects at a Distance

A telephoto lens for use with a camera is basically a type of telescope. The more usual kind of telescope consists of a long tube with a lens at each end. The front lens is the objective lens, and the rear one is the eyepiece. With two convex lenses the

TRY THIS

Camera in a Can

Lenses work by bending light rays. One of their best-known uses is in a camera. In this project, you will make a camera with no lens at all!

What to do

Wash an empty tin can thoroughly and dry it, being careful in case there is a sharp edge. Using a hammer and nail, punch a small hole in the center of the bottom of the can. You may want to ask an adult to help you. Cut a square of tissue paper larger than the diameter of the can, carefully put it over the open end of the can and secure it with a rubber band. Cut a piece of construction paper big enough to go all around the can and about 2 in (5 cm) longer than the can. Wrap it around the can and secure it with tape. Keep the paper flush with the closed end, but let it project beyond the tissue paper, forming a kind of hood.

With the hole away from you, point your can camera at a well lit object. The object will appear upside-down on the tissue paper "screen." The tiny hole acts like a lens, focusing rays of light from the object onto the screen. Just as in a real camera, the image is upside-down. Images formed in the backs of our eyes are also upside-down, but the brain turns them right side up.

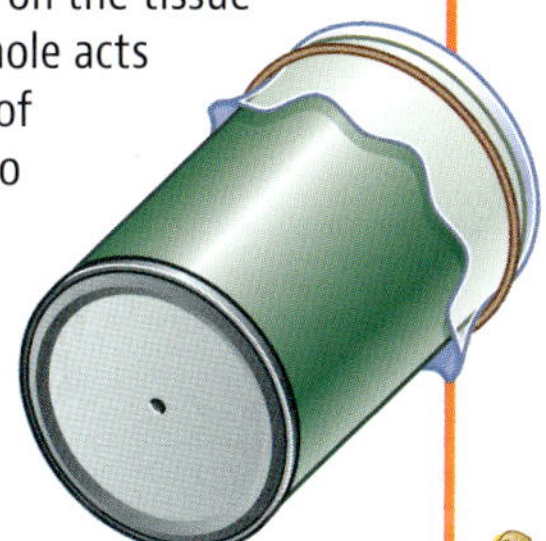

Make a hole in the base of the can, and stretch tissue paper over the open end.

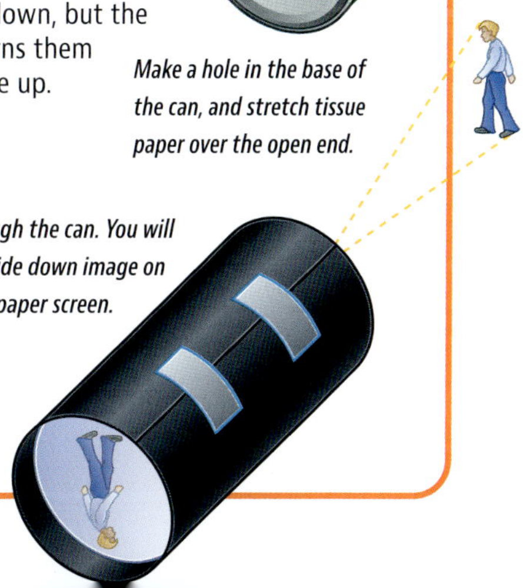

Look through the can. You will see an upside down image on the tissue paper screen.

REFLECTING TELESCOPES

Professional astronomers use reflecting telescopes to study the heavens. Light from distant objects first strikes a curved primary mirror, which brings the light to a focus, where it is viewed by an eyepiece. In most designs, the light is diverted to the focus by a smaller, secondary mirror. The Schmidt *camera*, invented in 1930, is used mainly for photographing large areas of the night sky. Its primary mirror is an easy-to-make spherical mirror; and to avoid spherical aberration, or irregularity, light first passes through a specially shaped corrector plate. An electronic light sensor is placed at the focus to record the image. The Newtonian reflector avoids spherical aberration by having a primary mirror with a cross-section in the form of a parabola—more steeply curved near the center than at the edges. Light focused by the primary mirror is directed to a focus at the side of the tube by a flat mirror at 45°. The Maksutov telescope has a corrector plate to overcome spherical aberration. The secondary mirror is a silvered spot on the back of the corrector plate, which reflects light back down the tube through a hole in the center of the primary mirror.

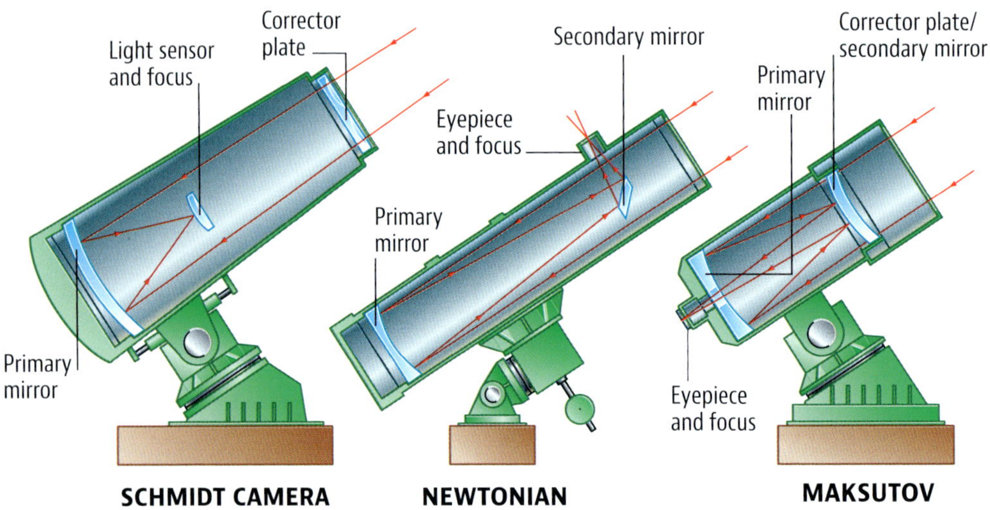

image is upside down, although this does not usually matter in astronomy. A third convex lens may be positioned inside the telescope tube, which will turn the image the correct way up. If the eyepiece is a concave lens, the image is the right way up.

This type of telescope is known as a Galilean telescope, after the Italian scientist Galileo Galilei (1564–1642), who used this same design in his pioneering astronomical studies nearly 400 years ago.

Binoculars

Paired telescopes are used in binoculars. Powerful telescopes are long and difficult to hold without a stand or a mounting, and they produce an inverted image. All of these difficulties are overcome in prismatic binoculars, in which each telescope has a pair of prisms that “fold” the optical path back and forth to shorten it. They are also arranged at right angles to each other, so that the final image is the right way up. Modern astronomers need really powerful telescopes. These types have curved mirrors instead of lenses.

Telescopes for Astronomy

Large mirrors are easier to make and are much lighter in weight than big lenses. There are various designs of mirror telescope, which are called reflecting telescopes or reflectors (see left); telescopes with lenses are known as refracting telescopes or refractors. Isaac Newton (1643–1727) made the first reflector. It had a single curved mirror at the bottom of a tube, with a small plane mirror angled near the other end. Light rays were reflected by the plane mirror to the eyepiece at the side of the telescope.

Most large, Earth-based telescopes used by professional astronomers have mirrors of up to 33 feet (10 m) across. Computer-controlled arms below the mirror adjust its shape to keep its curvature precise. Some very large mirrors are composed of many hexagonal panels, resembling a honeycomb. A computer controls the positions of the panels. Astronomical telescopes are mounted in large domed buildings, which can be rotated to aim the telescope at any part of the sky. The observatories are often located on high peaks, where the air is clearer, resulting in better images.

Very Small Objects

Microscopes produce enlarged images of very small objects. The first microscopes had a single, small convex lens. This type, called the simple microscope, was first made by a Dutch scientist named Anton van Leeuwenhoek (1632–1723) in about 1670. His best instruments

COMPOUND MICROSCOPE

This laboratory microscope has three lenses of different strengths. The stage, specimen platform, is moved up or down to focus the instrument using a coarse control for the main movement and the fine control for final adjustment. The condenser concentrates light onto the specimen. The yellow shading shows the path of light moving through the instrument.

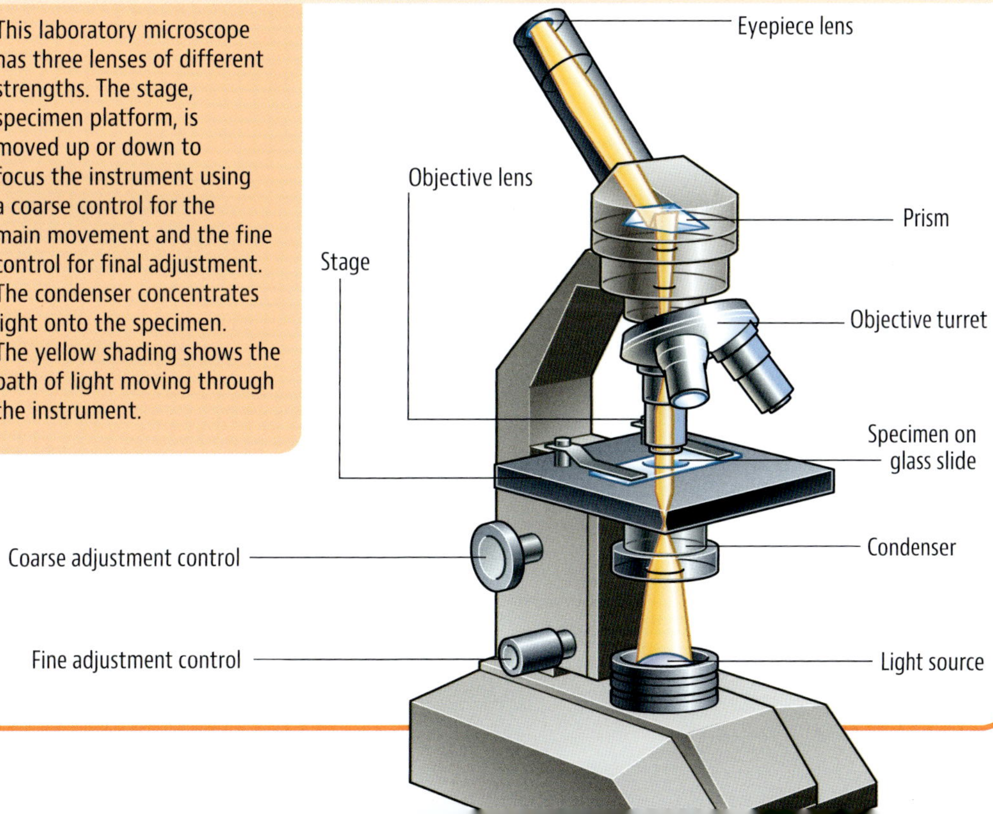

magnified more than 260 times. Van Leeuwenhoek was able to examine bacteria and blood cells using them.

To obtain higher magnifications, a compound microscope is used. It has a small but powerful objective (convex) lens. The image formed by the objective is further enlarged by the eyepiece lens, also convex. In use, the microscope is adjusted so that the object being examined is just beyond the focal length of the objective lens. This produces an enlarged inverted image inside the microscope tube. The image is just inside the focal length of the eyepiece lens, which acts like a magnifying glass to produce a greatly enlarged final image. The image remains inverted, but this seldom matters in most applications.

Most laboratory microscopes have a turret with two or three objectives of different magnifying powers. The specimen to be examined is placed on a glass slide mounted on a stage and illuminated from below by a bulb or by light reflected from a curved mirror. A pair of lenses, called a condenser, concentrates the light on the specimen. For some studies, for example the examination of geological rock specimens, a binocular microscope is used. This is a compound microscope equipped with two eyepieces.

SLIDE PROJECTOR

The optical system of a slide projector includes a curved mirror that acts as a reflector to direct light from a lamp onto the condenser. The condenser evens out the light, which then passes through the slide. Finally, a lens projects an enlarged image of the slide onto a distant screen.

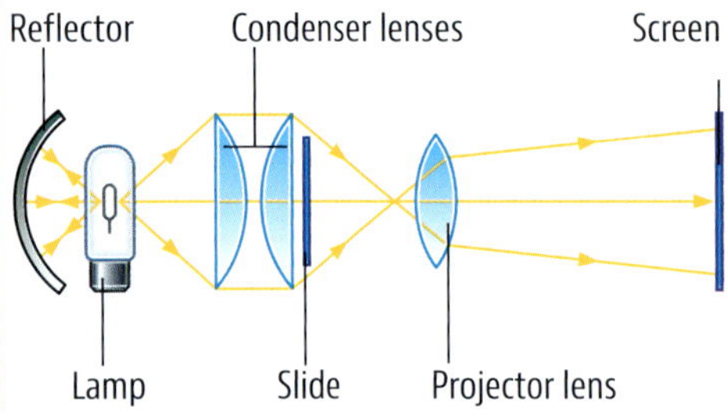

Projecting Images

One type of image that can be produced using film in an analog camera is a color transparency, commonly known as a slide. A slide can be viewed by holding it up to the light, but the image is very small and difficult to examine in detail. A better way of viewing it is to project an enlarged image of the slide onto a screen. This is the function of a slide projector.

In a slide projector, condenser lenses concentrate light from an electric lamp onto the slide and illuminate it evenly. Some condensers also incorporate a heat filter to avoid damaging the slide. The light passes through the slide, and the projector lens forms the image on the screen. The lens can be moved in or out slightly in order to focus the image. The image is actually upside down, which is why slides have to be inserted upside down into the projector.

Movies are typically projected digitally today, but old-style movie projectors are optically much the same as slide projectors. They have in addition a mechanism to move the film through the projector and a shutter that opens and closes rapidly (usually 24 times every second). Each individual picture, or frame, of the film is stationary when the shutter is open. Then, when the shutter closes, the film moves on to the next frame. With old-style movie projectors, we are actually seeing 24 still pictures each second but our brain ignores the brief black screens between consecutive images and simply "sees"

LASER LIGHT

Atoms in the ruby crystal emit light when they absorb energy from the flash tube. This light then stimulates more atoms to give off light, which bounces between the mirrors at the ends of the crystal. Coherent red laser light leaves through a hole in one of the mirrors.

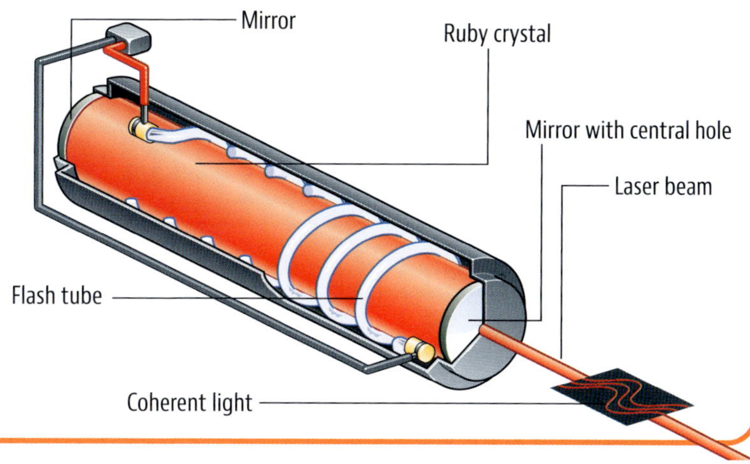

continuous movement. Likewise, most movie cameras today use digital video rather than film, but old-style movie cameras have a similar shutter arrangement, although in every other respect it is optically like the analog camera described earlier.

Lasers

Another different use of light is the laser, a device that produces an intense beam of light all of the same wavelength, with waves perfectly in step (coherent). "Laser" is an abbreviation of the term "light amplification by stimulated emission of radiation". The laser material can be a solid, such as a ruby crystal, or a gas, such as a mixture of helium and neon. Light energy from a flash tube energizes atoms of the laser material. Some of the atoms give off light, which stimulates more atoms to emit light in the same direction. Mirrors at each end of the laser bounce the light backward and forward, so that it gets increasingly stronger. The laser beam emerges from a central hole in one of the mirrors. Lasers have many different uses in medicine, industry, and communications. Small, microchip-based lasers are also used to scan groceries at store checkouts and in laser printers used in homes and offices.

An industrial laser cutting through a sheet of steel. Lasers cut with great precision. They are also used for welding.

SOUND WAVES

The air that surrounds us gives us more than life—it also gives us a world of sounds. Our nerves and brains convert those sounds into the experience of music, voices, and the everyday activities of the world around us.

Sound is a disturbance of the air. We are able to hear things because of the air between us and the things that cause sound. Sound is a particular kind of disturbance: it is a vibration. Not all disturbances of the air are vibrations. The wind is movement of the air from one place to another, revealed by the movement of clouds across the sky. The movement of air that carries a sound is not so obvious. It does not make the air move as a whole. Instead, tiny particles in the air called molecules vibrate individually—they move backward and then forward without any overall change in their position unless there is a wind blowing.

What Happens in a Wave?

Although we cannot see air, the vibrations of molecules in the air can be compared to the movement of water. Water moves from place to place in rivers and streams, in ocean currents, and in the flow from a faucet. These water movements are like the movement of air in winds. But waves in water are different. There can be waves even in calm water—on a lake, for example. These waves move along, but the water does not move along as a whole. A boat floating on the lake is not carried along; instead, it bobs up and down. As a wave passes beneath the boat, the individual molecules of water make a circular movement, going backward and forward and up and down. When the wave has passed, the molecules will have scarcely changed their positions, and neither will the boat. Sounds are waves in the air. A sound wave in still air travels on, while individual air molecules simply vibrate around a particular position.

In a sound wave, the molecules of air are crowded together, then separated. The movement is similar to the wave in a long, flexible spring, such as the toy springs that "walk" downstairs by themselves. Imagine the spring is hanging from one end. Jiggling the bottom end of the spring squeezes the coils of the wire closer together, with each coil pushing the next above it, and waves of compression move up the spring, separated by zones where the coils are momentarily stretched.

Air acts like a spring. When something vibrating, such as the skin of a drum, briefly "jiggles" it, waves of compression travel outward. In these zones of compression, molecules in the air crowd together. Between zones of compression the molecules are more thinly spread—the air is "rarefied." As the waves pass by, the molecules stay in the

This explosion is an uncontrolled release of energy that creates a devastating blast wave in the air. The blast consists of air rushing outward and traveling only a short way. Beyond the blast, sound waves are created that consist of air vibrating backward and forward. These waves travel long distances and are heard as an extremely loud bang.

same position. They vibrate backward and forward, first in the direction of the wave, and then in the opposite direction.

Sound, Noise, and Hearing

How we perceive a sound depends on our ears and brains. We cannot hear vibrations of the air that are very slow or very fast. We cannot hear vibrations in which the molecules make only very small movements; and vibrations that are very large hurt our ears and can even damage our hearing, perhaps permanently.

Sounds vary not only in their loudness and pitch, but also in their distinctive quality, or "color," which is called "timbre" (pronounced tam-ber). Notes played on a piano, a violin, a saxophone, and a flute sound quite different even when they all have the same pitch; the distinctive quality that enables us to recognize the sound of various instruments is the timbre. These differences all result from the complicated vibrations of the air molecules. Sounds from most instruments have a definite pitch—each note produced is distinct and different.

However, a mixture of sounds that have no definite pitch, such as in a large explosion, is called noise.

COMPRESSION WAVES

When sound radiates outward from a source, air does not move away from the source as a whole. Instead, the particles in the air, called molecules, vibrate backward and forward. Where the molecules are squeezed closer together, they are said to be compressed. Where they move apart, they are described as rarefied. The movement of the sound wave is the movement of this pattern. It is like a "Mexican wave" in a football stadium: spectators do not leave their seats, but stand and sit in succession so that a wave of motion runs through the crowd.

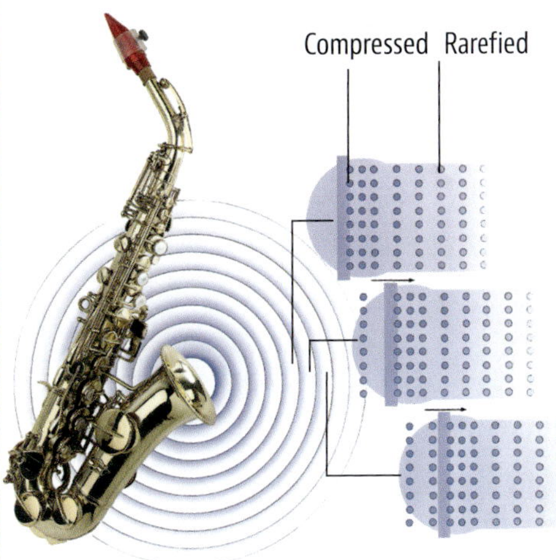

Area of compression moves forward

PROPERTIES OF SOUND WAVES

We can best understand the behavior of sound by thinking of it as a wave motion. Sound waves, like other waves, have certain important features, such as their wavelength and frequency, and the speed at which they travel.

To understand the behavior of sound waves, it is necessary to know something about waves in general. Water waves give the clearest illustration. The distance from one peak of a wave to the next (or from one trough to the next) is called the wavelength. In sound waves there are places, like the peaks and troughs of water waves, where the air molecules have shifted the most from their initial positions in the forward direction, and others where they have shifted the most in the backward direction. The distance from any one "peak" to the next, or from any one "trough" to the next, is the wavelength of the sound wave.

The number of waves passing by a given point every second is called the frequency of the wave. The frequency of the wave multiplied by its wavelength is the speed of the wave (speed = number of waves passing per second x the length of each wave). Under the same conditions all sound waves move at approximately the same speed. Higher-frequency waves have shorter wavelengths, and lower-frequency waves have longer wavelengths. Waves in water can be large or small. The height of a wave above the level of calm water is called its amplitude. The amplitude of a sound wave is the distance the air molecules move from their rest positions. The more violent the disturbance that causes the sound, the greater the amplitude and the louder the sound.

SCIENCE WORDS

Amplitude: The intensity of a wave. The amplitude of a sound wave is directly related to its loudness.

Decibel (dB): A unit of loudness. If a sound is 10 dB louder than another one, it is 10 times as intense and sounds roughly twice as loud.

Frequency: For a sound wave the number of times per second that the air molecules (or the molecules of whatever material the wave is traveling through) vibrate.

Wavelength: The distance between two successive locations where a wave is at its maximum intensity.

The Speed of Sound

Sound must have something to travel in, and it travels faster in some substances than in others. The speed of sound in air, for example, is about 1,080 feet (330 m) per second. As a result, it takes approximately five seconds for sound to travel 1 mile (1.6 km) or three seconds to travel 0.6 miles (1 km).

Sound travels faster in liquids than it does in air and other gases. In pure water, its speed is about 4,900 ft/s (1,500 m/s). It is slightly

The curved walls of the Whispering Gallery in London's St. Paul's Cathedral reflect sound waves so that someone whispering on one side of the gallery can be heard on the other.

faster in seawater. Sound travels fastest of all in solids. Its speed is about 16,500 ft/s (5,000 m/s) in steel, and in a hard rock such as granite it is about 19,700 ft/s (6,000 m/s). The sound of an approaching train can be heard as vibrations of the railroad lines before the sound of the locomotive reaches our ears through the air. Sound travels 15 times faster in steel than it does in air.

Decibels

Scientists measure loudness of sounds in decibels (symbol dB). The decibel equals one-tenth of another unit called the bel, named for Alexander Graham Bell (1847–1922), inventor of the telephone. A sound of 70 dB is 10 times as intense as a sound of 60 dB, but it sounds roughly twice as loud (not 10 times louder). The sound level in a busy city street is typically around 70 dB. In a disco or a rock concert the sound level is around 110 dB. A supersonic fighter taking off 550 yards (500 m) away puts out an ear-splitting 120 dB.

FREQUENCY, WAVELENGTH, AND AMPLITUDE

For any wave, the wavelength is the distance from one peak (or trough) to the next. The shorter the wavelength, the more waves pass a fixed point each second. Amplitude is related to the intensity, or strength, of a wave.

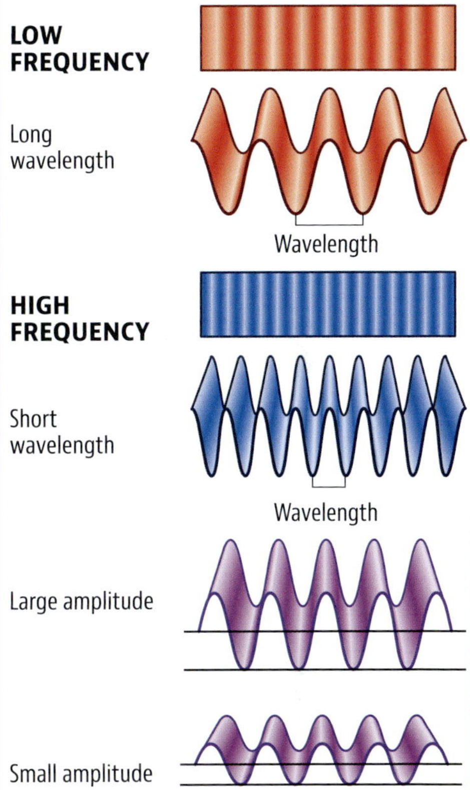

The loudness of a sound wave is related to its energy. This energy depends on the mass of air molecules set in motion by a vibrating object. The larger the mass of air, the louder the sound. The vibrating diaphragm in the earpiece of a telephone is too small to vibrate a large mass of air, so it cannot produce loud sounds. But the diaphragms in a rock band's loudspeakers can pump out more than 110 dB.

VIBRATING STRINGS

Stringed instruments have been around for several thousand years. Taut strings can be plucked, struck, or rubbed with bows to create sounds as varied as those of the harp and the violin. Strings can produce this difference in sound because it is possible for many waves to vibrate along the length of a single string at the same time.

The sound made by a vibrating string depends on its length, tension, and weight per unit length. The six strings of an electric guitar are all metal. The three with the highest pitch are single wires, the thinnest giving the highest pitch. The three lowest strings are made heavier by winding a second metal wire around a thicker steel wire core.

Many types of musical instrument have taut strings that vibrate when they are struck or plucked. Each stretched string vibrates with a frequency that is natural to it. This in turn sets the air in motion, making it vibrate at the same frequency as the string. The vibration dies away if the string is plucked or struck once. If the string is scraped with a bow, which is a set of taut horsehair or nylon strings stretched by a flexible wooden rod, the continued force keeps the string vibrating. The string makes a sustained note, but the note is still of the same frequency.

In any note produced in this way there are actually other, higher, frequencies mixed in with the main frequency, which is known as the fundamental. However, to understand how musicians produce different notes on stringed instruments, it is easier to begin by imagining that the fundamental is the only frequency present.

VIBRATING STRING

The simplest way in which a string can vibrate is when there are just two stationary points, called nodes, one at each of its ends. The maximum movement, called the antinode, is at the center. The vibration of the string sets up vibrations of the air, or sound waves, with the same frequency.

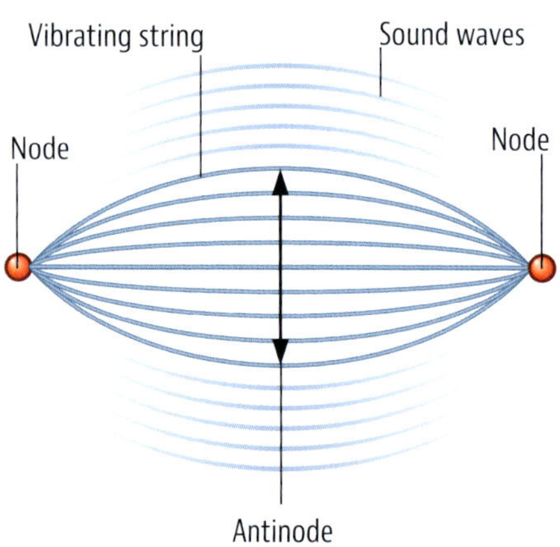

Changing Pitches

The frequency of vibration of a string depends on its tension. If you pull the string to make it more taut, the pitch rises. It falls again as the string is slackened. When a guitarist tunes up, you can hear the pitch of the notes from the strings rising and falling. Rock guitarists make use of this effect when they use the tremolo arm. It "wobbles" the bridge, the device on the body of the guitar where the strings are attached. This alters the tension slightly, making the pitches of the notes being played "wobble" too.

When the tension in a string is constant, the pitch of the note it gives out depends on its length. The shorter the string, the higher the note it produces. A harp has its characteristic shape because it is simply a set of strings, each with its own fixed length, providing a note in the range that is covered by the instrument. Players of bowed instruments vary the length of the strings by using their fingers to press the strings against the neck of the instrument—this is called "stopping"—at varying positions. Guitarists do the same, but there are ridges or bars called frets set into the neck of the guitar, providing a fixed endpoint for the string.

An early discovery of mathematical patterns in nature was made by the Greek philosopher Pythagoras (c. 570–c. 495 BCE). He found that the notes produced by a string depend on its length. If the length of the string is halved, the string produces a new, higher note that is harmonious with the original note. In fact, it is one octave above the original note. The sequence of eight notes in the major scale, do, re, me, fa, so, la, te, do, cover one octave. Harmonious notes are also produced when the string is stopped at two-thirds and four-fifths of its original length.

Another key feature of the strings in musical instruments is their weight. If two strings of the same length are stretched equally tightly, the heavier one will vibrate more slowly and give out a note of lower pitch. That is why the low-pitched strings of instruments are far thicker and heavier than the higher-pitched strings.

TRY THIS

Louder Twang

Here is a way of demonstrating that solids conduct sounds better than air does.

What to do

Stretch a rubber band between the finger and thumb of one hand and strum it with the fingers of the other hand. You will hear a faint twanging sound. Next, stretch the rubber band around a plastic glass (see the illustration below). Place the base of the glass against your ear, and strum the band again. This time, the twanging will be much louder. When a stretched string vibrates (on a violin or guitar, for example), it produces sound. When you strummed the rubber band stretched between your fingers, the sound traveled through the air to your ears. However, when you strummed the rubber band stretched around the plastic glass, the sound was carried to your ear through the plastic. Because plastic is a solid, it conducts sound much better than air does. The sound is also amplified a little by the space inside the glass, which acts rather like the hollow body of an acoustic guitar or violin.

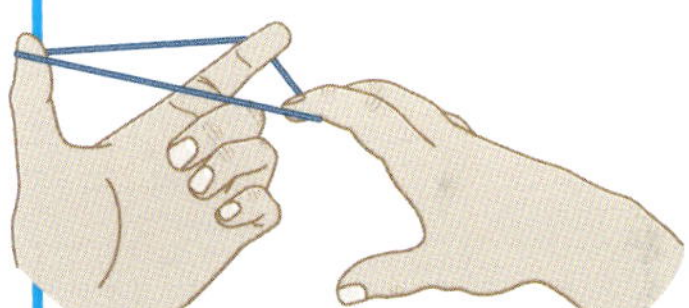

Stretch a rubber band and strum it. You can just hear a twanging sound.

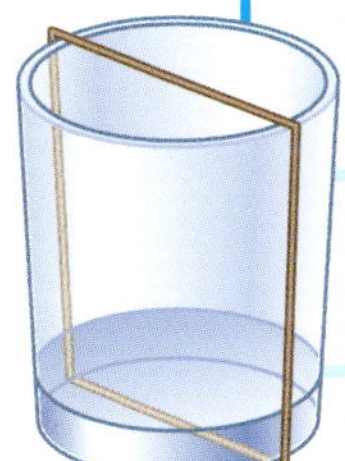

Stretch the rubber band around a plastic glass, hold the base of the glass to your ear and strum the rubber band.

VIBRATING COLUMNS OF AIR

Pipe organs, trumpets, flutes, and the human voice have something in common: they all make sound because a column of air vibrates. Unlike vibrating strings, which are struck, the amount of air that vibrates at each moment can be controlled by the player or speaker.

SCIENCE WORDS

Antinode: A position in a standing wave at which the amplitude is greatest. See also node.

Node: A position in a standing wave at which the amplitude is zero (or smallest). See also antinode.

Most of the sounds that we hear have been caused by something solid vibrating and setting up vibrations in the air. However, it is also possible for a quantity of air trapped inside some kind of enclosure to vibrate, causing vibrations in the air outside the enclosure that spread out freely as sound waves. An easy way to make this happen is to blow across the mouth of an empty bottle. When you blow at the correct angle, you will produce a loud, musical note that is made by the air inside the bottle vibrating. You can "tune" the bottle to make notes of a different pitch by adding some water. The new note will be higher than the first one because the amount of air inside the bottle is less when water is added. You an use a number of bottles tuned to different notes to play tunes.

The same principle is used in wind instruments. The air in the pipe or tube of the instrument vibrates to make the sound. The longer the column of air that vibrates, the deeper the note. Pipe organs consist of a set of pipes, each playing a single note. Air is blown into them mechanically. The organist uses hands and feet to press keys that allow air into whichever pipes are needed at each moment. In some organ pipes, called flue pipes, the air passes in through a specially shaped inlet that sets up the vibration. In others, called reed pipes, the air passes over a piece of flexible metal called a reed, and it is this that vibrates, setting the air in the pipe vibrating. Similar sorts of air inlet appear in other types of wind instrument.

One of the simplest of all wind instruments is the penny whistle. It is basically a tube with a shaped mouthpiece, called a fipple. The tube

VIBRATING AIRWAVES

The recorder and penny whistle are both members of the flute family. They are played by closing the holes with the fingers while blowing into the mouthpiece. Waves are set up with nodes at the closed holes, which determine the pitch of the note that is produced (here all the holes are closed).

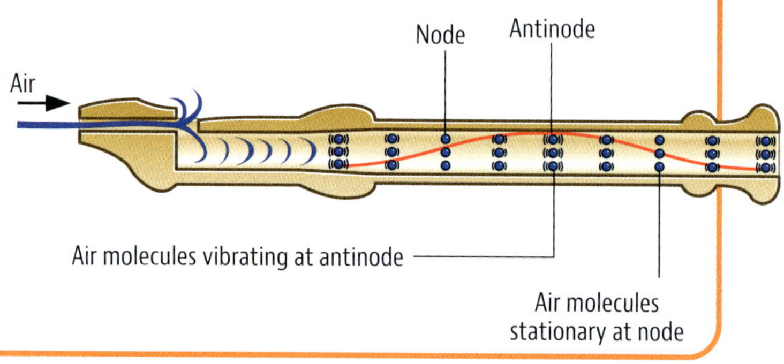

Brass players make their lips vibrate as they blow, which makes the air in the instrument vibrate. The length of the vibrating column of air is changed by pressing the valves down, which alter the effective length of the instrument's tubing.

is open at the other end, too. There is a series of holes along the pipe. By closing the holes with the fingers, the player alters the effective length of the pipe and changes the sound and pitch of the note.

Nodes and Antinodes

The pitch or frequency of the sound made by a vibrating column of air depends on its wavelength. The wavelength of a vibration is related to the distance from one node (a place where there is no vibration) to the next, or from one antinode (a place where there is maximum vibration) to the next. Opening and closing holes in a musical instrument alters the positions of antinodes, affecting the wavelength of the vibrations and the frequency of the sound.

Woodwind

The recorder and the flute are fundamentally the same kind of instrument as the penny whistle. Clarinets and saxophones differ in having a reed to provide the vibration; oboes and bassoons have a pair of reeds that vibrate against each other. All of these instruments are known as woodwind because originally they were made from wood.

TRY THIS

Pipes of Pan

A way of producing sound is to make a column of air vibrate inside a tube or a pipe. The pitch of the note—how high or low it sounds—depends on the width of the pipe and its length. In this project, you will make a set of musical pipes.

What to do

Stick several strips of double-sided tape to a piece of cardboard about 6 in (15 cm) square. Stick 12 drinking straws to the tape in a row, side by side, with the ends of the straws lined up along one edge of the card. Carefully cut through the other ends of the straws, and the cardboard, at an angle (see the illustration below). Hold the tops of the straws near your bottom lip and blow across and slightly downward. After a little practice, you will be able to get the straws to make sounds.

By blowing across the ends of the straws, you have made the air inside them vibrate. This produces the sounds. Notice that the short straws produce higher-pitched notes than the longer straws. The ancient Greeks made a musical instrument of graduated pipes, which they called a panpipe, after Pan, the god of flocks and fields. Native South Americans still play panpipes today.

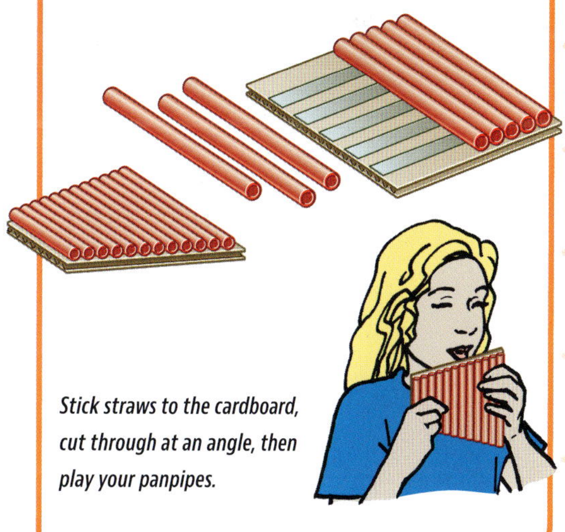

Stick straws to the cardboard, cut through at an angle, then play your panpipes.

VIBRATING SKINS AND SOLIDS

As well as using the human voice, ancient peoples first made music by knocking objects such as bones, stones, and pieces of wood together. Over time, sophisticated percussion instruments were developed. Percussion plays a major role in all forms of music worldwide

When you tap two pebbles together, you are unlikely to hear anything that is more interesting than a dull click. Hollow objects generally make louder and more musical sounds than solid ones. A wooden acoustic guitar makes a reasonably loud sound if you tap the body of the guitar with your fingers. This is because of the hollow body. If you cover up the sound hole and tap the body again, what you hear will be muffled and much quieter. A hollow log can make a booming noise. Dried seed-cases or gourds (a type of large, hollow fruit) make sharp, distinctive noises when tapped. These natural objects gave our prehistoric ancestors the idea of shaping solid objects to form musical instruments.

THE GLOCKENSPIEL

The glockenspiel consists of metal bars tuned a half-note apart. The metal bars are arranged like the keys of a piano, with the bars corresponding to the black piano keys in the upper row. The letters in the diagram are note names.

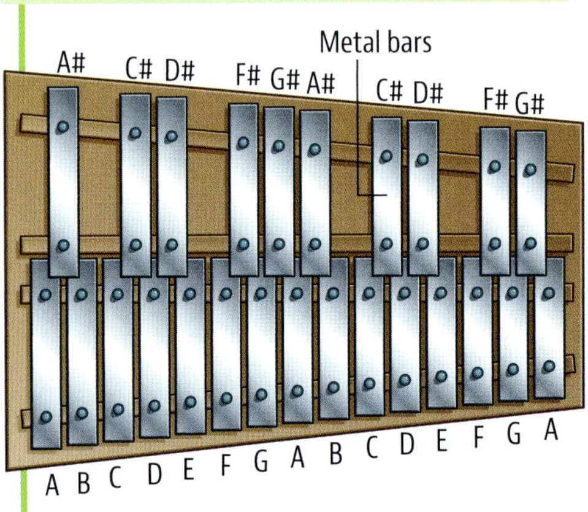

The Xylophone

A xylophone consists of wooden bars cut to precise lengths to give the desired notes when struck. Smaller wooden bars make higher notes. An array of these bars mounted on a frame makes an instrument with a wide range of frequencies. Hollow gourds, again of carefully chosen sizes, may be hung beneath the bars to amplify the sound. In the modern xylophone metal tubes are often used as

The bars of a vibraphone are made from metal. Tubular metal resonators are mounted beneath the bars to amplify the sound. Rotating disks within the resonators cause the pitch of each note to oscillate slightly.

resonators. Vibraphones have bars made from metal with tubular motor driven metal resonators mounted beneath the bars. They also have a sustain pedal, similar to a piano. This holds the note on for longer.

Drums

In the same way that a stretched string produces a musical note when it is plucked, so a taut skin or other membrane makes a note when it is struck. If the skin is stretched over

TRY THIS

Noisy Ruler

A xylophone is a percussion instrument that makes use of vibrating strips of wood. In this project, you will use a vibrating ruler to produce sounds.

What to do

Clamp the ruler to the table with your hand, leaving part of the ruler projecting over the edge. With your other hand, twang the free end of the ruler by pressing it down with a finger and releasing it suddenly. Try it several times with different lengths of ruler projecting. Which makes the highest note, a short length or a long length?

For a different sound, press the ruler against the table with most of its length projecting over the edge. Twang it as before, but quickly slide the ruler farther back onto the table.

The pitch of the note depends on the length of ruler that is vibrating. The shorter the length, the higher the note. When you slid the vibrating ruler farther onto the table, the length was changed from long to short. As a result, the note changed pitch from low to high.

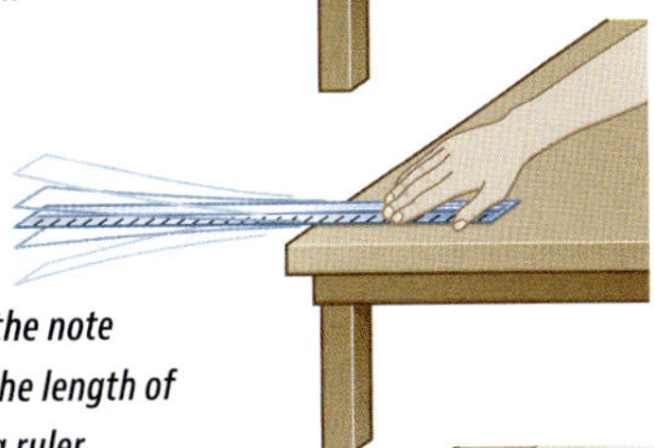

The pitch of the note depends on the length of the twanging ruler.

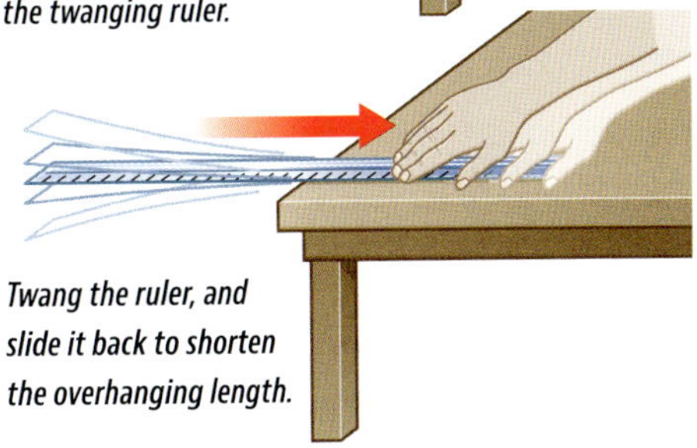

Twang the ruler, and slide it back to shorten the overhanging length.

TRY THIS

Ringing Spoons

Vibrating objects produce sounds, and sounds travel through materials. In this project, you will make ordinary spoons chime like church bells.

What to do

Tie the handle of a teaspoon at the center of a piece of string about 4 ft (1.3 m) long. Wind the ends of the string around your index fingers, keeping both halves of the string the same length. Place the tips of your fingers in your ears (do not push them in too far!). Lean forward so that the spoon hangs free, and let it hit the leg of a chair or a table (or ask a friend to hit the teaspoon with a wooden spoon). You will hear a chiming sound like a church bell. Repeat the procedure with larger spoons. What happens to the pitch of the sounds—are they higher or lower?

The sounds come from the vibrating spoons, and the string carries the sounds right into your ears. The larger the spoon, the lower the tone of its chime. You can make the sounds even louder by passing the ends of the string through holes in the bases of paper or Styrofoam cups. Tie a knot on the inside of the cup, or tape the string into place and put the open ends of the cups over your ears.

A vibrating spoon can make a chiming sound like distant church bells.

a hollow vessel, a loud, pleasing noise is produced. In many types of drum there is a mixture of many frequencies in the note, which does not have the pure sound of a plucked string. Some drums, such as the steel drums made from oil containers, are designed to be tuned precisely.

Keeping Up the Tension

The note made by a drum depends not only on its size, but also on the tension—the tautness—of the skin. The tighter the skin, the higher the note. One traditional type of drum used in Africa and elsewhere has skin stretched over both ends by cords. The drum body has a "waist" so that the strings are not touching the body. The player holds the drum under one arm and squeezes the cage of cords while beating the drum with the other hand, creating a varying pitch.

A symphony orchestra usually has four or more tunable drums called kettledrums (or timpani). Originally the drum's skin was tightened (to raise the pitch of the note) or

PERCUSSION INSTRUMENTS

Some of the huge family of instruments that can be found in the percussion section of an orchestra.

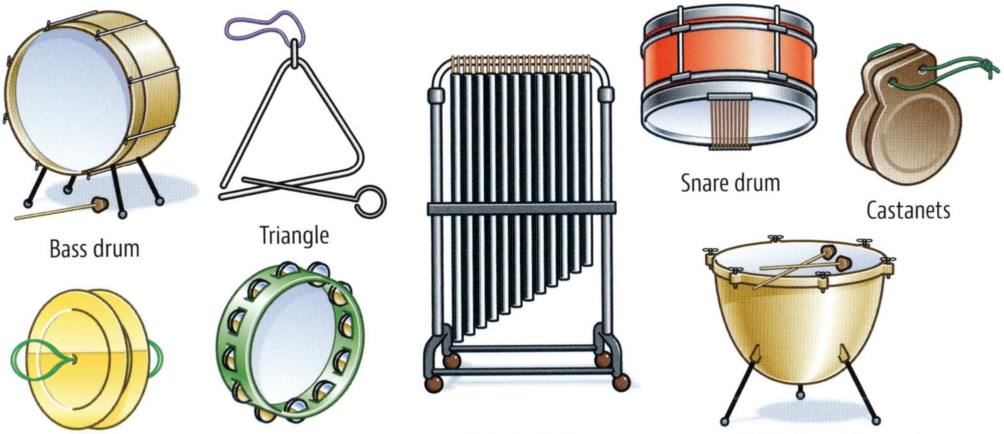

A drummer playing a drum kit consisting of several types of drums and cymbals. Their sounds fall into different ranges, and they can have different qualities or "colors" of sound according to the way in which they are struck.

slackened (to lower the note) by turning keys mounted at intervals around the rim of the drum. In more modern kettledrums a foot pedal is used to stretch or slacken the skin.

Bells, Gongs, and Cymbals

Metals make a satisfying clang when struck. Their polycrystalline structure built from lattices of atoms vibrate with precise frequencies when they are hit. In the past, maintenance engineers checked the condition of wheels on trains by hitting them with a hammer. A good wheel produced a musical note, but a cracked one made a dull, flat sound. Most church bells are made from metal, usually bronze.

Gongs are usually cast or beaten out of bronze. Unlike a bell, a gong does not have a definite pitch, but it produces a mixture of tones. A cymbal is similar, but is made from a disk of brass. Cymbals are often played in pairs to produce a metallic crashing sound when they are struck together.

SPEED OF SOUND

Sound travels at a definite speed, but the speed varies according to the medium in which the sound travels. Sound waves can be outpaced by jet planes, by space rockets during launch, and even by jet-propelled cars.

Picture yourself in the middle of a thunderstorm. You see a lightning stroke 3 miles (5 km) away. The light reaches you almost instantly—after about 1/60,000 of a second. But the sound, traveling at a speed of 1 mile (1.6 km) every 5 seconds, or 0.6 miles (1 km) every 3 seconds, takes 15 seconds to arrive.

There is even a delay over shorter distances. The sound of a hammer blow 330 feet (100 m) away is delayed by a noticeable one-third of a second from the sight of the hammer falling. However, you never see this in movies. The boom of a distant explosion usually coincides with the sight of it. Film directors think their audience will be puzzled by a delay!

SCIENCE WORDS

Doppler effect: The change in frequency of a wave caused by the relative motion of the source and the observer. For example, the pitch of an ambulance siren seems to fall as the vehicle passes and goes away from a listener.

Mach number: The ratio of the speed of an object to the speed of sound.

Shock wave: A disturbance that moves through a fluid, such as air or water, faster than the speed of sound in that fluid.

Sonic boom: An explosive sound caused by the shock wave preceding an aircraft traveling at or above the speed of sound.

The speed of sound in air increases with increases in temperature. At 32°F (0°C) it is 1,088 feet (331.6 m) per second in dry air. At 68°F (20°C) it is 1,129 ft/s (344 m/s). The temperature of the atmosphere decreases up to an altitude of about 8 miles (13 km), and so does the speed of sound. At this height the speed of sound falls to about 938 ft/s (286 m/s).

Mach Number

The ratio of the speed of an object to the speed of sound under the same circumstances is called its Mach number. Mach 1 is the speed of sound, Mach 2 is twice the speed of sound, and so on. The number is named for the Austrian physicist Ernst Mach (1838–1916).

A jet fighter of the U.S. Navy breaking the sound barrier. Objects traveling faster than the speed of sound form a conically shaped high pressure surface, seen around this airplane, and generate a shock wave and sonic boom. This cone is caused by the sound source (the jet) traveling faster than the sound waves produced by it. High humidity levels make the cone visible.

Sound waves travel faster through liquids and solids. At sea, a distant explosion will be heard twice—the first sound travels through the water, the second sound has traveled through the air. In water, the speed of sound is about four and a half times its speed in air. That is why it is hard to judge the distance of something like a far-off speedboat when you are swimming in the sea or underwater.

DOPPLER EFFECT

As a police car speeds toward a bystander with its siren blaring, each sound wave has a shorter distance to travel than the previous one. The frequency is raised, and the sound rises in pitch. When the car has passed and is speeding away, the sound waves are stretched, their frequency is lowered, and the pitch of the siren falls.

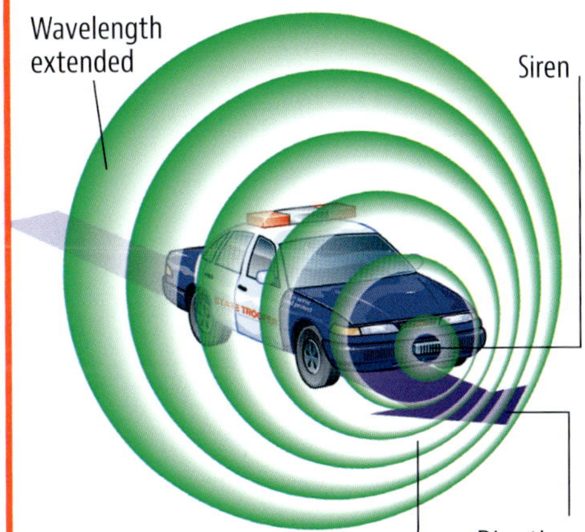

CHRISTIAN DOPPLER

The Austrian physicist Christian Doppler (1803–1853) showed how apparent frequency and wavelength vary according to the motion of the source and the listener.

When the source of the waves is approaching the listener, or vice versa, the apparent frequency increases. If the source and the listener are moving apart, the apparent frequency is lowered. In 1845, Doppler ran an experiment to demonstrate his theory. A locomotive pulled a carriage of trumpeters past other musicians, who judged the change in pitch of the notes as the train went past. Their judgments matched Doppler's predictions.

ULTRASONIC AND SUBSONIC VIBRATIONS

We are surrounded by ultrasound—sound so high-pitched that we cannot hear it. But animals can hear and make these sounds, and respond to them. They fill the air and oceans with unheard cries. People make use of ultrasound in industry, warfare, and medicine.

The deepest sounds humans can hear consist of about 20 vibrations per second, or 20 hertz—named for the German physicist Heinrich Hertz (1857–1894). The symbol for hertz is Hz. The highest audible frequency is about 20,000 Hz, or 20 kilohertz (1 kilohertz, symbol kHz, is 1,000 hertz).

Sound waves with frequencies higher than this are called "ultrasonic." Many animals are able to hear this ultrasound. Dog whistles produce loud ultrasound, which dogs react to but which people cannot hear. Bats have an uncanny ability to find their way and hunt prey in darkness. They give out high-pitched squeaks at ultrasonic frequencies that can be as high as 200 kHz, and they detect the echoes from insects and other objects.

Producing Ultrasound

The usual way of producing audible sound is through a loudspeaker, in which a paper or metal diaphragm vibrates rapidly to create the sounds. But a diaphragm cannot be made to vibrate fast enough to produce ultrasound. Instead, the sound producer—called a transducer—employs a crystal that is made to vibrate by an oscillating electric current applied across its faces. The crystal is usually made from quartz or a chemical called Rochelle salt. Ultrasound produced in this way is used for cleaning things. Clothing can be submerged in water or cleaning fluid and rapidly agitated to remove dirt by beaming in an ultrasound signal.

The bats' method of navigation is used by seafarers. Ultrasonic pulses sent out from ships are reflected from, for example, submarines, shoals of fish, or the seabed. The time the echo takes to return to the ship shows the distance of the object. A similar method is used in medicine. High-frequency pulses penetrate the body. Echoes from internal organs show the body's interior.

Subsonic Sound

The deep, low-frequency sounds from a church organ, or from a truck or plane, can make a whole building throb. In these cases,

SCIENCE WORDS

Loudspeaker: A device that produces sound from an electrical signal.
Ultrasound: Sound of frequency too high to be heard by human beings.

Ultrasound can be used to look inside the body. This type of scan is routine for women during pregnancy.

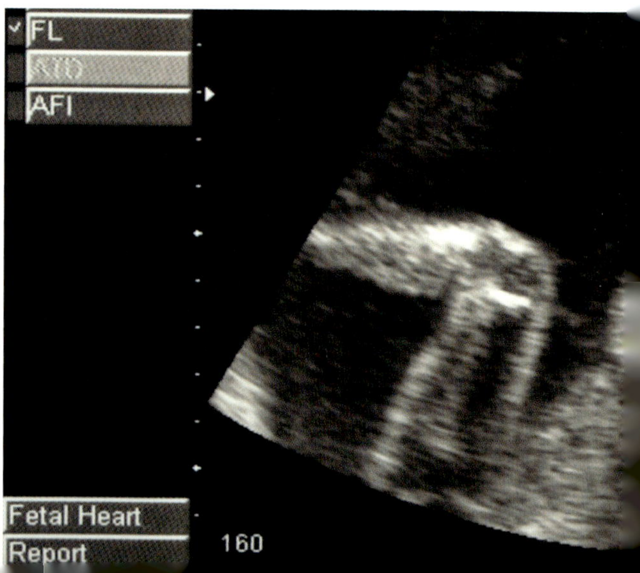

as well as the sounds we can hear there are vibrations that are of lower frequency. These vibrations are called infrasound, or subsonic sound. Although human beings cannot hear infrasound, many animals can, among them whales and elephants.

Lower-frequency sounds travel farther than higher-frequency ones. That is why distant thunder is always deep and rumbling, while a nearby thunderclap is much sharper, with a cracking tone to it. Some animals take advantage of the long range of low-frequency sound. Whales emit infrasound to communicate over distances of hundreds of thousands of miles/km.

In the 1980s, naturalists discovered that elephants produce subsonic rumbling noises from their nasal passages. They use these sounds to communicate with one another over long distances.

HUNTING WITH ULTRASOUND

Bats navigate using echolocation/ultrasound. Most of the sound they produce is too high-pitched for us to hear. The bats' large and sensitive ears detect echoes, enabling the bats to home in on insects and avoid obstacles. So amazingly accurate is their navigational ability that bats can fly safely through darkened rooms criss-crossed by wires.

Sound and Movement

The throbbing that a deep sound makes is caused by a phenomenon called resonance. An object is said to resonate when it is made to vibrate by the vibration of some other nearby object. For example, sounding a single note near a piano—by plucking a guitar string, say—will cause the piano string that is tuned to the same note to vibrate. Other strings, such as those an octave higher or lower, will do the same. The vibrations of the air from the plucked string tend to make all the piano strings vibrate, but only those piano strings that naturally vibrate at the same frequency as the guitar string will develop a strong vibration.

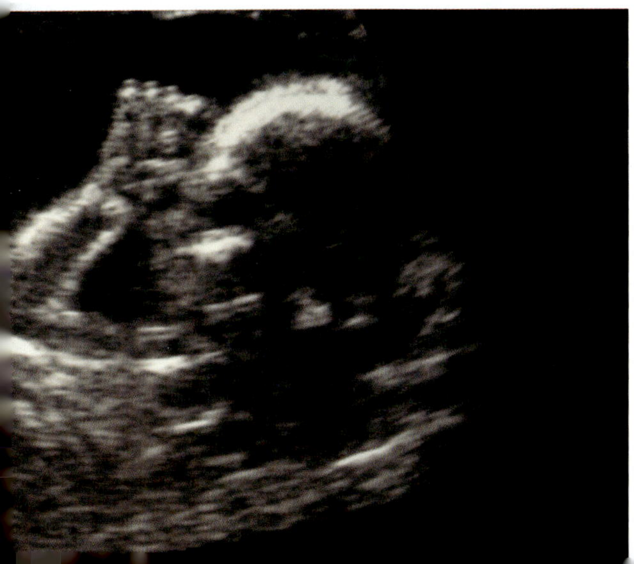

In a similar way, if someone pushing a child on a playground swing gives small pushes at the same rate the child is swinging, the movement of the swing will build up. Pushes at a different rate will be out of step with the regular motion of the swing, and motion will be prevented from building up.

Buildings and furniture do not usually resonate to normal sound frequencies. Because the natural rates of vibration of these objects are low, they resonate (if they resonate at all) to infrasound. Over time a concert hall or church can be damaged by deep notes from an organ. A spectacular example of harmful resonance occurred in 1940, when the wind caused the Tacoma Narrows suspension bridge in Washington to vibrate uncontrollably. The deck of the bridge snaked like a shaken rope, flinging cars into the river below and finally collapsing. Since that disaster new bridges have always incorporated antiresonance features.

The Shaking Earth

Most earthquakes are caused by shifts within the top 190 miles (300 km) of Earth's mantle. The shifts are small, but affect enormous quantities of rock. The shifts cause vibrations of all frequencies in the surrounding crust. Some vibrations travel for large distances through the crust, others through the deeper parts of Earth—the mantle and core.

Sound waves in a solid, unlike sound waves in a fluid such as water or air, can be of two kinds. In one, atoms and molecules are made to vibrate backward and forward in the same direction as the motion of the wave. This is the same as the movement of the particles in a sound wave in a gas or liquid. But in the other kind of sound wave in a solid, the atoms or molecules move from side to

RICHTER SCALE

Until relatively recently, the magnitude of earthquakes have been measured on the Richter scale. The amount of damage caused by the earthquake depends not only on its magnitude on the Richter scale, but also on its depth and on the numbers of people and buildings in the area. Today, geologists use the Moment Magnitude Scale, which uses the same logarithmic scale, but measures the strength of larger earthquakes much more accurately.

2.5 Not generally felt, but recorded on seismometers
3.5 Felt by many people
5.0 Some local damage
7.5 Destructive earthquake
8.5 Major earthquake

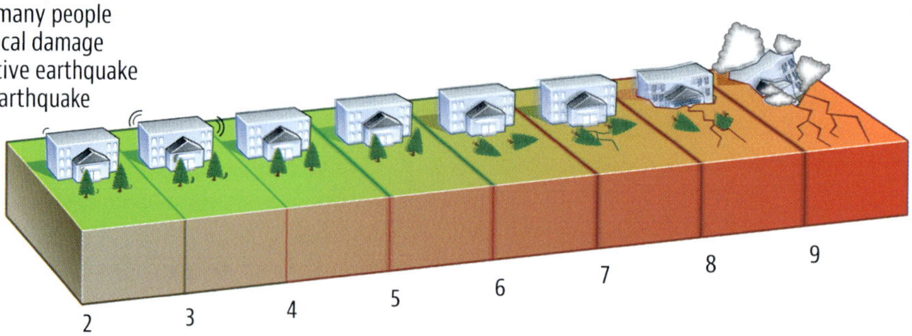

This sort of damage can be caused by an earthquake that shakes the ground for only a few moments. Earthquake vibrations are the sound of a planet being shaken.

side. These waves are like the waves you can send along a rope that is fixed at one end by shaking its free end.

Shear Waves and Pressure Waves

In the study of earthquakes, the side-to-side waves are called S (shear) waves, and the faster, pressure waves are called P waves. The different conditions of pressure, density, and rock composition at different depths below the surface of Earth affect the speeds of both S waves and P waves.

The varying conditions in the body of the planet cause the paths of earthquake waves (known as seismic waves) to bend. The pattern of waves arriving at different places around the world following an earthquake is very complex, but over many years they have been disentangled to reveal a lot about the structure of Earth. They reveal a difference between the lighter rocks of the thin crust and the denser rocks of the mantle, which extends halfway toward the center of the Earth. And they show that there is a liquid core of molten iron and nickel that the S waves cannot penetrate. There is also an innermost solid iron–nickel core.

When a tremor is felt in a particular area, the ground generally moves both from side to side and up and down. Seismographs, or Earth-tremor measuring devices, record these tremors as graphs. Traditionally, the seismograph consisted of a pendulum with a heavy weight, or bob, that was free either to bounce up and down on a spring or to swing from side to side. The bob carried a pen that made a trace on paper mounted on a revolving drum.

During a tremor the heavy bob tended to remain still, while the rest of the instrument shook, producing a wavy trace on the paper. However, today most seismometers work electromagnetically, with data recorded electronically on computers or recorded onto paper to create a seismogram.

CHARLES RICHTER

U.S. seismologist Charles Richter (1900–1985) will always be associated with the earthquake scale named after him. Richter developed it with Beno Gutenberg (1889–1960), another seismologist, in 1935. They rated the amount of energy released in an earthquake by the size of the marks made by the pen of a seismograph, with a correction for the distance of the earthquake from the observing laboratory. The points on the scale do not represent equal intervals. Going from one point on the scale to the next, the energy released in the earthquake is multiplied by about 30.

REFLECTION AND REFRACTION OF SOUND

Even in total darkness we can sense the presence of objects around us. This ability is largely due to our subconscious perceptions of sounds reflected by objects. The principle of reflection of sound is used in sonar to detect objects under water.

Sound waves, like all other types of wave, can be reflected. They can be reflected by solid objects, by the surface of water, and even by the boundary between layers of water that are at different temperatures beneath the surface. A reflected sound is called an echo. An echo from a distant object is noticeably delayed compared with the original sound. When you shout or clap your hands to hear the echo, every 5 seconds' delay represents 1 mile (1.6 km) and a delay of 3 seconds represents 0.6 miles (1 km). To figure out the distance of the object reflecting the sound, the delay time has to be divided by two, because the sound has traveled both ways by the time the echo is heard.

Echoes All Around

Echoes from nearby objects are not obvious, but they do affect the quality of all the sounds we hear. A radio program made in the open air sounds quite different from one made indoors, where there are echoes from walls and ceilings. The sounds in a huge cathedral are very different from those in a small kitchen. A room echoes when it is empty of people and of soft furnishings, such as drapes and upholstered chairs, which absorb a great deal of the sound. The characteristics of the sound inside

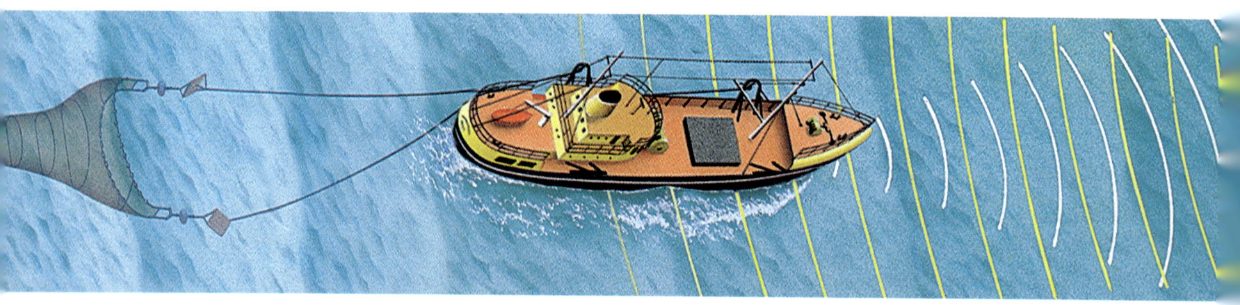

SCIENCE WORDS

Acoustics:
1) The science of sound.
2) The sound qualities of, for example a room, theater, or concert hall.

Echo: A sound that is reflected from some surface or object before reaching a listener, and so is delayed in relation to the original sound, which arrives directly.

Echolocation: Finding the direction and distance of an object by beaming sound pulses at it and detecting the echoes. The technology used is called sonar.

A military helicopter lowering a side-scanning sonar buoy into the ocean. Devices such as this send out pulses that bounce off the sea bottom and can detect hazardous underwater obstacles like wrecks, or the presence of explosive mines.

a room or building are called its acoustics. Multiple echoes can be very confusing. For example, somebody listening to a speaker but sitting toward the back of a large hall hears sounds by several routes, but not all at exactly the same time. In addition to the sound of the speaker's voice arriving directly, sounds also arrive a fraction of a second later, having been reflected from the walls and from the ceiling. This mixture of sounds distorts the speaker's voice. Such reflections can be avoided by covering the walls and ceilings with sound-absorbent materials, such as tiles made from foamed plastics.

In a well-designed theater or concert hall the tip-up seats are specially manufactured so that each one absorbs about the same amount of sound as they would if a person was sitting in it. As a result, the hall will have the same acoustic properties even if many of the seats are empty.

Refraction of Sound

The paths of sound waves can be gently bent as well as sharply reflected. Sound waves, like

Sonar is of huge importance to the fishing industry. Even small fishing vessels search for shoals of fish using sonar. From the trace on the sonar screen, an experienced operator can judge the size of the shoal and can often tell what type of fish it is.

MAKING ECHOES

You hear an echo of your own voice when some of its sound energy is reflected from an object and returned to you. A rocky cliff or the wall of a building make good echoes.

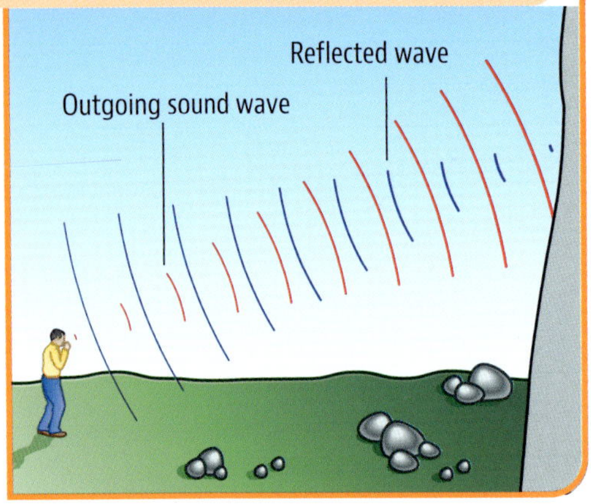

other waves, usually bend when their speed changes. Sound slows down when it enters cooler air. Normally, air high up is cooler than air near the ground. Sound waves rising at an angle to the horizontal from a source on the ground are slowed as they rise. The braking effect drags them around so that they are bent upward. This is rather like a column of soldiers marching at an angle onto rough ground, which slows them down. They tend to deviate from the direction they were marching in. This bending upward reduces the loudness of the sound at ground level some distance from the source.

In the open air, sounds are weaker at a distance than they would be in air of uniform temperature. The effects can be the reverse when air is warmer higher up than it is near the ground. This happens in certain weather conditions and is common in the polar regions, where the air is chilled near the frozen ground. Sound waves are speeded up as they rise upward, and this makes them swerve downward. It is then often possible to hear quiet sounds over great distances.

Banks of loudspeakers at coastguard stations used to send powerful blasts of sound as a warning to ships when there was fog. The shape of the loudspeaker horns helped to concentrate sound waves into a narrow beam. Maritime foghorns have been largely superseded by more modern navigational and safety technology such as radar and GPS.

Blowing in the Wind

The wind also affects the audibility of sound for a similar reason. Windspeed normally increases with height, and the effect of this is to make any sound waves traveling upwind bend upward, and any sound waves traveling downwind bend downward. This makes it easier for people to hear other people speaking, for example, when they are standing upwind of you (see illustration, top right). Sounds can be made easier to hear if they are focused

into a beam. Beams of sound are stronger within the beam and weaker anywhere outside the beam. A loudspeaker, an old-fashioned phonograph, and a loudhailer all have a horn with a shape that is designed to reflect the sound waves into a limited beam, rather than spreading them outward in all directions.

SCIENCE WORDS

Amplifier: A device that increases the strength of a sound or of an electronic signal representing a sound, as in a radio or hi-fi.
Beam: A concentrated stream of particles or a similar configuration of light or sound waves.
Downwind: In the same direction toward which the wind is blowing, that is, with the wind from behind.
Upwind: In or toward the direction from which the wind is blowing.

HEARING IN THE WIND

Windspeed usually increases with height. Higher above the ground, sound waves going upwind are slowed down. All the sound waves going in that direction are therefore bent upward. They are hard to hear at ground level. Higher-altitude sound waves going downwind are speeded up and bent downward, making them easier to hear.

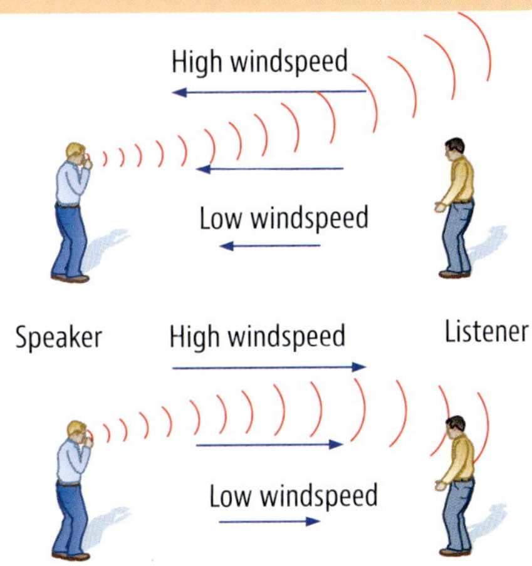

HORN OF PLENTY

The earliest types of record-players, or phonographs, did not have electronic amplifiers or loudspeakers. Sound was generated by the vibration of a needle as it moved in the record groove. A large horn amplified the sound toward the listener.

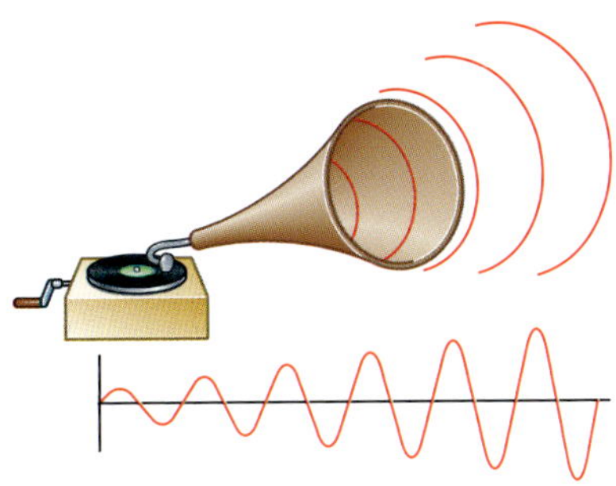

THE HUMAN EAR

Our ears allow us to make sense of the world of sounds around us. The analysis of sounds, telling us what frequencies are present in them, takes place in hidden structures within our head. Our sense of balance is located in the same structures. Most defects of hearing, apart from total deafness, can be corrected by physicians.

We can keep our balance, sense which way up we are, and judge how our body is moving thanks to the semicircular canals, which are organs in the inner ear.

Human beings detect sound by means of a complex sensory mechanism. The visible flaps of skin that we call ears, and anatomists call pinnas or pinnae, are only the outermost part of a structure that goes deep into our head. This structure is divided into the outer, middle, and inner ears.

Sounds from the outside world pass down a passage called the auditory canal and make the eardrum vibrate These vibrations are passed through a chain of three small bones to a complicated fluid-filled structure in the inner ear called the cochlea. The vibrations of the fluid in turn cause small, fine cells called hair cells to vibrate, generating electrical signals. The signals travel to the brain, and a sound is perceived. The deepest, longest-wavelength sounds travel farthest along the cochlea; so when hairs in that far part of the cochlea vibrate, the brain can "work out" that there are low frequencies in the sound that has been heard.

SCIENCE WORDS

Auditory canal: The passage that runs from the visible ear, or pinna, into the head.
Cochlea: A spiral organ in the inner ear that contains fluid and has many nerves attached. It is responsible for the detection of sounds.
Eardrum: Also called tympanum, a membrane separating the outer ear from the middle ear. It vibrates when sound waves strike it, passing the vibrations to a chain of small linked bones in the middle ear.
Inner ear: The part of the ear that contains the organs of hearing and balance. See also Cochlea; Semicircular canals.
Semicircular canals: Loop-shaped, fluid-filled organs in the inner ear that provide the sense of balance.

Achieving Balance

Our sense of balance is also controlled from the inner ear, where there are three fluid-filled loops called the semicircular canals at right angles to one another. Movements of the head set the fluid in the canals in motion, providing information from which the brain can compute the position and movement of the head in relation to the direction of gravity. Giddiness is caused by the fluid in the canals continuing to move after the head has stopped moving.

Hearing Problems

There are various things that can upset the delicate balance of the hearing mechanism. These can have results that vary from hardness of hearing to profound deafness. The simplest cause of partial deafness is an obstruction in the auditory canal, most often by earwax, which is easily removed. Inflammation of the middle ear (otitis) is another possible cause, which can also usually be treated successfully. More difficult to remedy are problems with the ear's sound-detecting apparatus, the auditory nerve, or the brain's hearing center. A hearing aid may help; an alternative is a cochlear implant, in which audio signals are fed to an electrode placed in the inner ear.

Tuning forks like the one being used here by a piano tuner give out a pure tone of a single frequency. Such purity of sound is almost unknown in nature.

INSIDE THE EAR

The pinna, or ear flap, is shaped to hear the direction from which a sound comes. Sound waves travel along the auditory canal, making the eardrum vibrate. In the middle ear, which is connected to the nose and throat, the vibrations of the eardrum are passed along three bones joined together, called the hammer (malleus), anvil (incus), and stirrup (stapes). The stirrup sets up vibrations in the fluid-filled inner ear. Hairlike cells in the coiled tube of the cochlea send electrical signals to the brain. The three fluid-filled semicircular canals detect head movements.

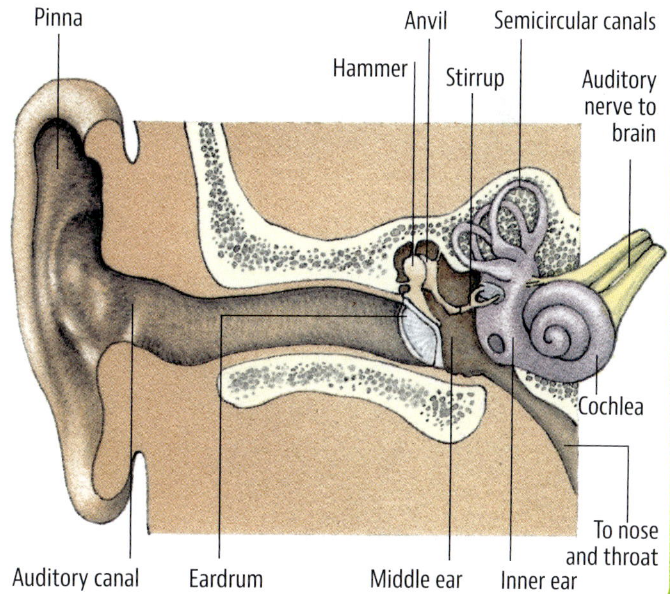

THE HUMAN VOICE

The human voice is complex and flexible. Subtle, controlled movements of the jaws, tongue, lips, teeth, and vocal cords together produce the countless different types of vocal sounds that can be made. Children master this control in the first few years of life, but scientists are still patiently uncovering the intricacies of human speech.

Whenever we shout, speak, sing, or whistle our vocal cords—which are relaxed when we are simply breathing—come together.

The human voice works like a combined wind and stringed instrument. Air from the lungs blows between two organs called the vocal cords, which are housed in the larynx, or voicebox. This lies at the front of the throat and at the top of the windpipe, or trachea (see diagram opposite).

The vocal cords form a V shape across the windpipe. They are made from elastic tissue and can be pulled by muscles in the larynx. When the muscles are relaxed, the vocal cords are relatively far apart, and no sound is made by air passing through. The muscles can move the cords together, and then they vibrate as air passes between them. The muscles can also tighten the cords, raising the pitch, or loosen them, to lower the pitch. At puberty there are changes in the larynx. The vocal cords of boys generally become less tightly stretched, so the pitch of the voice drops.

The epiglottis is part of the larynx. This flap of cartilage closes off the top of the windpipe when we swallow so that food cannot enter it and perhaps choke us. The food is diverted into the esophagus, or gullet, the tube that

SHAPING SOUNDS

These are mouth movements that produce some of the consonant sounds. B and P, for example, are called plosive because the lips are pressed together and then parted, releasing air explosively. T and D are called alveolar, because the tongue is briefly pressed against the alveolus, the front of the roof of the mouth. Other sounds have different classifications.

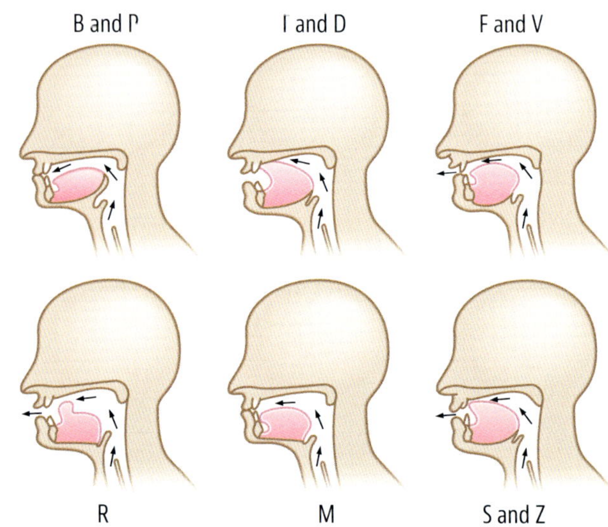

leads to the stomach. Some sounds are made without the vocal cords vibrating. They are called unvoiced, and include the P and T sounds. The B and D sounds, by contrast, are made with vibration of the vocal cords and are described as voiced. Voiced sounds also include all the vowels.

The sounds generated in the larynx are modified in the mouth. We make complex movements of the tongue, lips, and teeth in order to form the vowel and consonant sounds of speech. Tongue and lip positions are shown for some consonants in the diagram below left.

There is a wide range of frequencies in human speech. The ability to detect high frequencies is important in distinguishing many of the consonants from one another. The commonest forms of deafness that develop as people get older involve difficulty in distinguishing sounds in this way.

Voiceprints

There are individual differences in the mix of frequencies in different people's voices, even when they are speaking the same words. These frequencies can be electronically analyzed and represented in an image called a voiceprint.

A voiceprint can be used to identify a speaker in a similar way to how a fingerprint can be used to identify an individual. A recording of a nuisance telephone call can be turned into a voiceprint that can then be compared with the voiceprint of a suspect.

SCIENCE WORDS

Larynx: Also called the voicebox, the organ in the human throat that contains the vocal cords.
Vocal cords: The elastic organs in the throat that produce the human voice by vibrating as air is expelled between them.

MACHINERY OF SPEECH

The vocal cords lie in the larynx, at the top of the windpipe, or trachea, at the front of the throat. When speaking, a person blows air through the vocal cords while altering their position and tension to vary the pitch of the sounds produced. Movements of the mouth make further alterations in the sounds to produce recognizable speech.

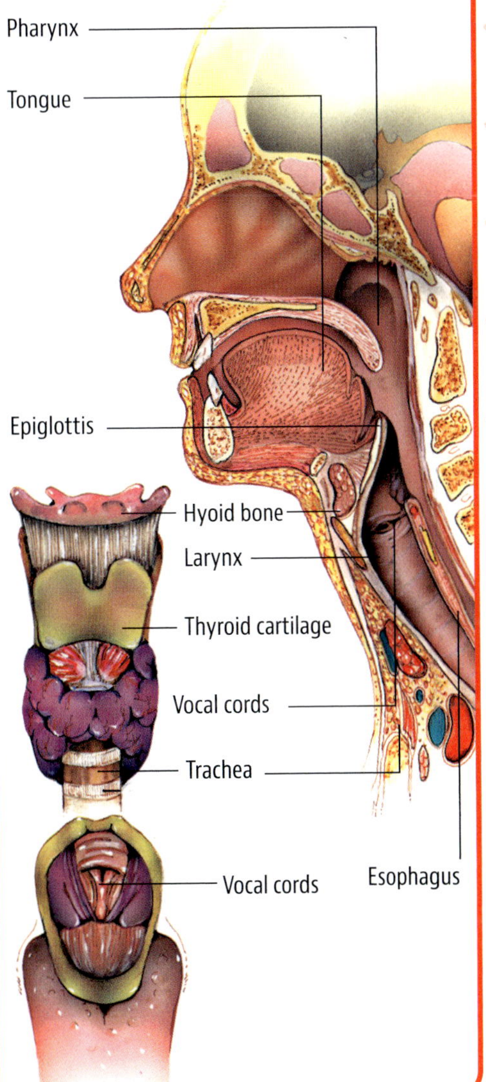

SOUND RECORDING AND REPRODUCTION

Nothing seems more elusive than a sound, which vanishes in a moment. Yet we have learned to store sounds, send them around the world, make them loud or soft at will, and mix or remix them into entirely new forms. With modern technology, the sounds made by performers are now just the raw materials.

The First Sound Recordings consisted of grooves in a wax surface. The shape of the groove was a direct imitation of the shape of waves of sound. These first recordings were made in 1877 by the American inventor Thomas Alva Edison (1847–1931) on a phonograph. A person would speak or sing loudly in front of the open end of a large horn, and the sound was focused to make a metal plate vibrate. A needle attached to the plate cut a wavy groove into the soft surface of a revolving wax-coated cylinder. To play back the sound, the same equipment was used in reverse. The cylinder rotated, making a needle in the groove vibrate. The needle was attached to a plate, which also vibrated, reproducing a faint version of the original sound, which was amplified by a large horn.

Disks were soon discovered to be far more convenient than cylinders. The first 78 rpm (revolutions per minute) records were made from a type of resin called shellac. Vinyl, a type of plastic, was used for 33 and 45 rpm records from the 1950s onward. Vinyl records reached the height of their popularity in the

From the 1980s, miniaturization made it possible for people to listen to music on personal compact cassette decks, Walkmans, MiniDiscs, and MIDI. By the 21st century, recordings of music and other audio files were digital. Music apps and iTunes turned smartphones into portable music centers.

MAGNETIC TAPE RECORDING

During recording, the erase head uses magnetic fields to remove any signal already on the tape. The record head then magnetizes the tape in a pattern that corresponds to the original sound.

When played back, the twin tracks of magnetization cause electrical signals to be generated in the read head, one for each stereo channel.

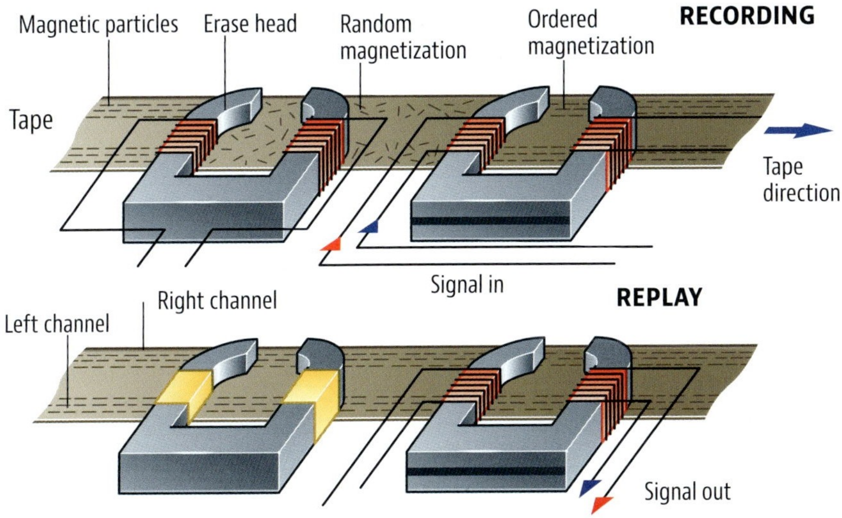

1960s and 1970s, but fell from favor when music met computing in the 1990s. Compact disks (CDs) replaced cassettes and vinyl. However, from about 2007 vinyl made a comeback, with supporters claiming that analog disks give better sound quality than digital devices.

Electronic Recording

Sound recording in the early days was much improved when electronic microphones and loudspeakers were developed. Electronic microphones generate electric currents varying in strength, representing the varying intensity (loudness) of the sound from moment to moment. This signal controls the cutting of the original disk. When a vinyl disk is played back, the vibrations of the stylus (pickup head) are converted into an electric signal, which is used to drive a loudspeaker.

Magnetic Recording

Recording on vinyl disks achieved great quality, but was difficult to do outside the studio. Magnetic tape recording could be carried out to a reasonable standard by amateurs, and the equipment eventually became portable and convenient. A magnetic recording tape consists of a plastic film coated with metallic particles. The particles become magnetized when exposed to a strong magnetic field. During recording, the tape is wound past two electromagnets. An electromagnet consists of a wire carrying an electric current wrapped around a metal core. The current creates a magnetic field, which the metal core increases. So the whole device is a controllable magnet.

The first electromagnet in a tape recorder is called the erase head, and its job is to remove any existing magnetization on the tape. The second electromagnet is called the record/replay head. A varying current flows through it, representing the sound being recorded. This creates a varying magnetic field, and it in turn magnetizes the metal coating on the tape. A pattern of strong and weak magnetization is put onto the tape.

During playback the tape is wound past the record/playback head. The changing magnetic field generates a changing electrical signal in the head, which is converted into sound.

Compact Disks

The vinyl disk is an analog device: the wavy shape of the groove is just like a picture of the sound wave. Compact disks and digital versatile disks (CDs and DVDs) are digital devices that store sound information in the form of a sequence of numbers (digits). The disks are plastic-coated aluminum disk in which tracks consisting of tiny pits about 24 millionths of an inch (0.6 of a micrometer) across spiral outward from the center. At each position along the track a pit may be present or may be lacking. These two possibilities represent the two digits 0 and 1.

Combinations of two digits are enough to represent any number in binary code. Music recorded on tape or vinyl disk contains sounds such as echoes and hisses. No unwanted sounds are included on CD or DVD, making the sound reproduction clearer than it had ever been before.

The MP3 digital file was introduced in 1996. MP3s only record the parts of the music that can be detected by the human ear. Because of the small file size, MP3s can be easily downloaded and emailed. They changed the way people bought and listened to music.

COMPACT DISK

The underside of a CD is etched with billions of tiny pits, arranged in a spiral track. There are also regions, called flats, where there are no pits. The CD player's read head swings between the center and the edge of the disk as the disk spins. A laser beam (red) is shone onto the disk, and its reflections are sensed. The long sequence of pits and flats is thus translated into a stream of electronic signals. They, in turn, are converted into sound in the loudspeakers.

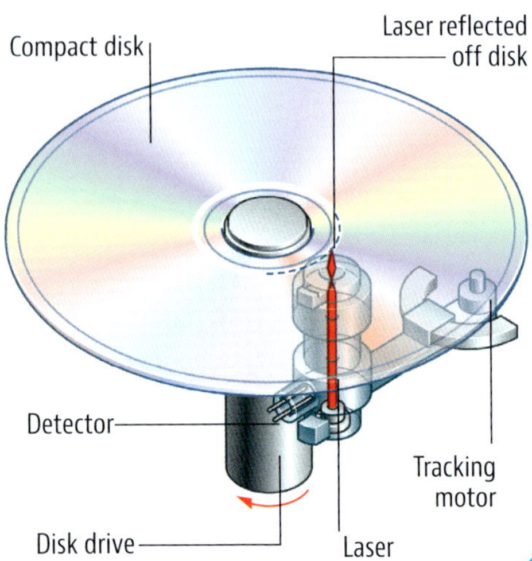

SCIENCE WORDS

Analog: Describing a signal or display that continuously varies. An analog signal resembles what it represents: for example, the groove on a traditional vinyl record is like a picture of the changing amplitude of the sound. See also Digital.
CD: Abbreviation for compact disk, a medium for recording sound, images, and data. It consists of a metal-coated plastic disk on which digital information is recorded as a track consisting of tiny pits.
Digital: Describing a signal or display that uses digits (numerals) to represent something. The display on a digital watch represents time as numbers rather than by the movement of hands. See also Analog.
Magnetic tape: A plastic tape coated with a magnetic material that can record sounds and other forms of information.

Today, professional digital music recording software is available for free download. Recording artists play, create, record, sample, download, edit, and store tracks and sounds directly onto their computers for immediate upload to the Internet or other devices.

TRY THIS

Paper Amplifier

The first phonographs had a large metal horn to amplify the sound from the record. People also used to use horns—called megaphones—to amplify their voices. In this project, you will make a horn and use it to listen to an old record.

What to do

Cut two squares of strong paper, one about 6 in (15 cm) square, and the other twice as large. Form each piece of paper into a cone by rolling it from one corner and taping it together. Carefully push a steel pin crosswise through each cone near the pointed end so that the point of the pin sticks out. Put an old, unwanted record on the turntable and turn it on. Hold a cone lightly while placing the point of the pin in a groove on the record. Can you hear anything? Repeat with the other cone. You should have heard whatever sounds were on the record. The wavy groove in the record vibrates the pin, which then vibrates the paper horn. The horn vibration causes the air inside it to vibrate and produce sounds, exactly in time with the sound vibrations on the record. The larger horn vibrates more air and so produces louder sounds.

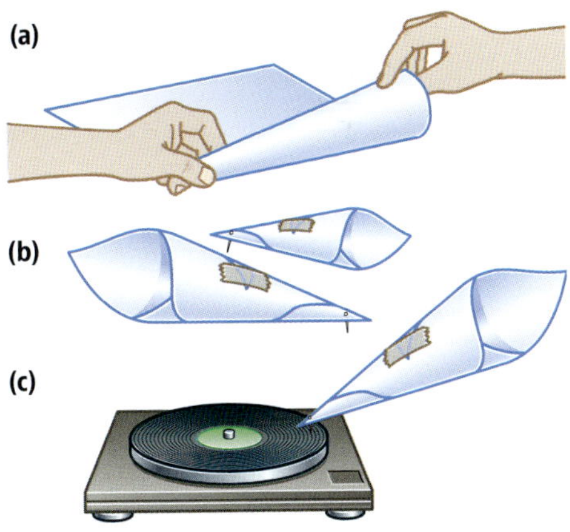

Roll the horns **(a)**, *tape them and add a pin* **(b)**, *and place the pin in the record groove* **(c)**.

GLOSSARY

Acoustics 1. The science of sound. 2. The sound qualities of, for example, a room, theater, or concert hall.

Amplifier A device that increases the strength of a sound or of an electronic signal representing a sound, as in a radio or hi-fi.

Amplitude The intensity of a wave. The amplitude of a sound wave is directly related to its loudness.

Analog Non digital. Describing a signal or display that continuously varies. An analog signal resembles what it represents. For example, the groove on a vinyl record is like a picture of the changing amplitude of the sound.

Angle of incidence The angle between an incident ray and the normal to a mirror or to the surface of a block of transparent material.

Angle of reflection The angle between the reflected ray and the normal to a mirror.

CD Abbreviation for compact disk, a medium for recording sound, images, and data.

Concave lens Also called a diverging lens, a type of lens that causes parallel rays of light to spread out (diverge) as if coming from a point (the focus) behind the lens. Its surfaces curve inward.

Concave mirror Also called a converging mirror, a type of mirror that causes parallel rays of light to be reflected to a focus in front of the mirror. Its surface curves inward.

Convex lens Also called a converging lens, a type of lens that causes parallel rays of light to converge to a point (the focus) in front of the lens. Its surfaces curve outward.

Convex mirror Also called a diverging mirror, a type of mirror that causes parallel rays of light to spread out (diverge) after reflection so that they appear to come from a point (the focus) behind the mirror. Its surface curves outward.

Decibel (dB) A unit of loudness. If a sound is 10 dB louder than another one, it is 10 times as intense and sounds roughly twice as loud.

Digital Describing a signal or display that uses digits (numerals) to represent something. For example, the display on a digital watch represents time as numbers rather than by the movement of hands.

Dispersion The splitting of white light into the colors of the rainbow (a spectrum), for example, by a triangular prism.

Doppler effect The change in frequency of a wave caused by the relative motion of the source and the observer.

Echolocation Finding the direction and distance of an object by beaming sound pulses at it and detecting the echoes. The technology used is called sonar.

Electron A subatomic particle with a negative electric charge. Electrons surround the nucleus of an atom. They play a key role in electricity, magnetism, and the conduction of heat.

Fluorescent bulb Also called a fluorescent tube, an electric lamp consisting of a tube containing mercury vapor, with electrodes at each end.

Frequency For a sound wave, the number of times per second that the air molecules (or the molecules of whatever material the wave is traveling through) vibrate.

Gravity The natural force of attraction exerted by the Earth, which draws bodies at or near its surface toward its center.

Incandescence The emission of light by an object that is heated to white heat.

Incandescent lamp An electric bulb that has a filament (usually made from tungsten) in a glass globe containing traces of an inert gas such as argon. The electric current heats the filament to incandescence.

Laws of reflection of light 1. The angle of incidence equals the angle of reflection. 2. The incident ray, the normal, and the reflected ray all lie in the same plane.

Lens A piece of transparent material that bends light, changing the direction of light rays passing through it.

Lunar eclipse An eclipse of the Moon, occurring when the Earth's shadow (cast by the Sun) falls onto the Moon.

Mach number The ratio of the speed of an object to the speed of sound.

Magnetic tape A plastic tape coated with a magnetic material that can record sounds and other forms of information. The strength of magnetization of the tape at each point represents the loudness of the sound at a particular moment.

Node A position in a standing wave at which the amplitude is zero (or smallest).

Photoelectric cell Also called a photocell, a current-producing device consisting of an element such as silicon that emits electrons when struck by light.

Plane A flat surface, for example, things on the same plane could all be drawn on a flat sheet of paper.

Prism A block of glass or plastic that can bend light or split it into the colors of the rainbow. Often triangular in shape.

Reflected ray A ray of light that is reflected by a mirror.

Refracted ray A ray of light that is refracted as it passes from one transparent material into another.

Refraction The bending of light rays as they pass from one transparent material into another.

Semisilvered A mirror having a metallic film backing of such thinness that only half the incident light is reflected, the remainder being transmitted through the backing.

Shock wave A disturbance that moves through a fluid, such as air or water, faster than the speed of sound in that fluid.

Solar eclipse An eclipse of the Sun, caused by the Moon passing between the Earth and the Sun.

Solar panel A device that converts light from the Sun into electricity and heat. A solar panel is made from a collection of many small solar cells over a large area that work together to produce large amounts of power from sunlight.

Supersonic flow Flow of a fluid at more than the speed of sound in that fluid. It is marked by shock waves.

Ultrasound Sound of frequency too high to be heard by human beings.

Vacuum A completely empty space in which there are no atoms or molecules.

Wavelength The distance between two successive locations where a wave is at its maximum intensity.

FURTHER RESEARCH

Books

Baker, Laura. *Physics for Curious Kids.* Arcturus, 2022.

Colson, Rob. *Spectacular Light and Sound (Extreme Science).* New York, NY: Gareth Stevens Publishing, 2023.

Stuart, Colin. *The Speed of Starlight: How Physics, Light, and Sound Work.* Big Picture Press, 2020.

Web Sites

Explore Sound: Acoustical Society of America
https://exploresound.org/
Have fun learning about the science of sound, including experiments.

How Stuff Works
https://science.howstuffworks.com/light2.htm
Topics on all aspects of light and how it works.

Optics for Kids
https://www.optics4kids.org/
Learn about optics, the science of light.

INDEX